Focus On MIDDLE SCHOOL

3rd Edition

Rebecca W. Keller, PhD

Real Science-4-Kids

Illustrations: Janet Moneymaker

Copyright © 2019 Rebecca Woodbury, Ph.D.

All rights reserved. No part of this publication may be reproduced, stored in a retrieval system, or transmitted, in any form or by any means, electronic, mechanical, photocopying, recording, or otherwise, without prior written permission from the publisher. No part of this book may be reproduced in any manner whatsoever without written permission.

Focus On Middle School Astronomy Student Textbook—3rd Edition (softcover)
ISBN 978-1-941181-45-4

Published by Gravitas Publications Inc.
www.gravitaspublications.com
www.realscience4kids.com

Contents

CHAPTER 1 WHAT IS ASTRONOMY? — 1

    1.1 Introduction — 2
    1.2 Early Astronomers — 2
    1.3 Modern Astronomers — 4
    1.4 Changing Views of the Cosmos — 4
    1.5 Summary — 7
    1.6 Some Things to Think About — 7

CHAPTER 2 TECHNOLOGY IN ASTRONOMY — 8

    2.1 Introduction — 9
    2.2 Telescopes — 10
    2.3 Space Telescopes and Other Satellites — 12
    2.4 Other Space Tools — 14
    2.5 Summary — 17
    2.6 Some Things to Think About — 18

CHAPTER 3 EARTH IN SPACE — 19

    3.1 Introduction — 20
    3.2 The Earth in Space — 20
    3.3 The Earth and the Moon — 22
    3.4 The Earth and the Sun — 25
    3.5 Eclipses — 27
    3.6 Summary — 28
    3.7 Some Things to Think About — 28

CHAPTER 4 THE MOON AND THE SUN — 29

    4.1 Introduction — 30
    4.2 The Moon — 30
    4.3 The Sun — 33
    4.4 Chemistry and Physics of Stars — 34
    4.5 Summary — 36
    4.6 Some Things to Think About — 36

CHAPTER 5 PLANETS — 37

- 5.1 Introduction — 38
- 5.2 Types of Planets — 38
- 5.3 Earth-like Planets — 39
- 5.4 Jupiter-like Planets — 41
- 5.5 What Happened to Pluto? — 44
- 5.6 Summary — 45
- 5.7 Some Things to Think About — 46

CHAPTER 6 TIME, CLOCKS, AND THE STARS — 47

- 6.1 Introduction — 48
- 6.2 Reading a Star Atlas — 50
- 6.3 Time — 52
- 6.4 Celestial Clocks — 55
- 6.5 Summary — 57
- 6.6 Some Things to Think About — 57

CHAPTER 7 OUR SOLAR SYSTEM — 58

- 7.1 Introduction — 59
- 7.2 Planetary Position — 59
- 7.3 Planetary Orbits — 60
- 7.4 Asteroids, Meteorites, and Comets — 62
- 7.5 Habitable Earth — 65
- 7.6 Summary — 66
- 7.7 Some Things to Think About — 67

CHAPTER 8 OTHER SOLAR SYSTEMS — 68

- 8.1 Introduction — 69
- 8.2 Closest Stars — 69
- 8.3 Brightest and Largest Stars — 71
- 8.4 Planets Near Other Stars — 72
- 8.5 The Circumstellar Habitable Zone — 74
- 8.6 Summary — 76
- 8.7 Some Things to Think About — 77

CHAPTER 9 GALAXIES — 78

- 9.1 Introduction — 79
- 9.2 Discovering Galaxies — 79
- 9.3 Clusters — 83
- 9.4 At the Galactic Center — 84
- 9.5 Star Formation — 86
- 9.6 Galaxies Interact — 86
- 9.7 Summary — 89
- 9.8 Some Things to Think About — 90

CHAPTER 10 OUR GALAXY—THE MILKY WAY — 91

- 10.1 Introduction — 92
- 10.2 Shape and Structure — 93
- 10.3 Size — 97
- 10.4 Viewing the Milky Way Galaxy — 99
- 10.5 Summary — 103
- 10.6 Some Things to Think About — 103

CHAPTER 11 OTHER GALAXIES — 104

- 11.1 Introduction — 105
- 11.2 Spiral Galaxies — 106
- 11.3 Barred Spiral Galaxies — 107
- 11.4 Elliptical Galaxies — 109
- 11.5 Irregular or Peculiar? — 110
- 11.6 Radio Galaxies — 112
- 11.7 Summary — 114
- 11.8 Some Things to Think About — 114

CHAPTER 12 EXPLODING STARS AND OTHER STUFF — 116

- 12.1 Introduction — 117
- 12.2 Red Giants, White Dwarfs, and Novae — 117
- 12.3 A White Dwarf Goes Supernova — 123
- 12.4 Supergiant Stars and Supernovae — 124
- 12.5 Summary — 128
- 12.6 Some Things to Think About — 128

GLOSSARY-INDEX — 130

Chapter 1 What Is Astronomy?

1.1	Introduction	2
1.2	Early Astronomers	2
1.3	Modern Astronomers	4
1.4	Changing Views of the Cosmos	4
1.5	Summary	7
1.7	Some Things to Think About	7

1.1 Introduction

Astronomy is considered by many to be the oldest science. Since long before the invention of the telescope, human beings have been looking at the stars. The word astronomy comes from the Greek words *aster* which means "star" and *nomas* which means "to assign, distribute, or arrange." The word astronomy literally means "to assign or arrange the stars." Astronomers are scientists who assign names to all the celestial bodies in space, including stars, and study how they exist and move in space.

1.2 Early Astronomers

The earliest recorded history reveals an interest in the stars. Cave drawings show primitive humans recording observations from the skies, and later the Babylonians recorded detailed planetary positions, eclipses, and other astronomical observations. Egyptian and Greek observers expanded on the information collected by the Babylonians. Some people think that the pyramids in Egypt align with the stars of Orion and this suggests that the Egyptians acquired sophisticated abilities to observe the sky. The Ancient Greeks were the first astronomers to add mathematics to astronomy.

Many early civilizations used the stars and the movements of celestial bodies as tools to measure time. The Sumerians of Babylonia used the phases of the Moon to create the first lunar calendar, and the Egyptians, Greeks, and Romans copied and revised this calendar. Today our calendar is derived directly from the Sumerian calendar and is connected to the monthly and yearly orbits of the Moon and Earth.

On the other side of the ocean, the Incan and Mayan civilizations created sophisticated calendars by observing the planetary cycles. The Mayan calendar is circular and has aspects that relate the movement of the Sun, Moon, and planets.

Early astronomers named individual stars as well as groups of stars that form constellations. A constellation is any group of stars that fit together to form a pattern in the night sky. Some of the major constellations that come from Greek mythology are familiar to many people.

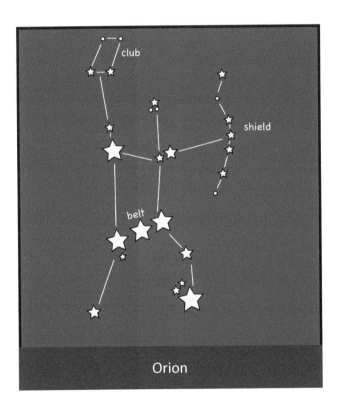

Orion

Orion the Hunter is a constellation of stars that can be seen from the Northern Hemisphere from December through March. Orion has a "belt" of three bright stars in a straight row. Once the "belt" is located, it is easy to find the "club" and "shield" by looking for neighboring stars.

The constellation names derived from Greek mythology have changed very little since 1000 BCE. There are currently 88 constellations that are recognized by the International Astronomical Union (IAU), and over half of those were observed by the ancient Greeks!

1.3 Modern Astronomers

Today, astronomers can see many more stars than their ancient predecessors could. Modern astronomers can also see details about the planets and stars that were not visible in ancient times.

Telescopes, radios, and cameras are just some of the tools astronomers use when studying the planets and stars. Modern astronomers also use chemistry and physics to understand astronomical data. Understanding how planets move requires knowing the physics behind gravity, inertia, and mass. Understanding how stars give off heat and light energy requires knowing the chemistry behind nuclear reactions. And understanding how the Sun affects our weather requires knowledge of magnetic and electric fields. Modern astronomers not only have sophisticated tools to explore the universe, they also have centuries of complicated mathematics, chemistry, and physics to help them understand how the universe works.

1.4 Changing Views of the Cosmos

The practice of astronomy changed dramatically after the invention of the telescope, a scientific tool that uses lenses to magnify distant objects. In the 1600's Galileo Galilei, an Italian scientist considered to be the first modern astronomer, used the telescope to look at the planets. Galileo was also able to see the moons of Jupiter and the rotation of the Sun. Based on his observations, Galileo confirmed a radical new view of the cosmos. The cosmos, or solar system, includes our Sun and the planets around it.

In ancient times most people believed that the Earth was the center of the universe. These ancients believed that the planets and the Sun moved in a circular orbit, or path, around the Earth. This view of the world is called geocentric. Geo comes from the Greek word *ge* which means "earth" or "land" and centric comes from the Greek word *kentron* which means "point" or "center." A geocentric view is one that considers the Earth as the true center of the universe.

Geocentric Cosmos

It is not hard to understand why this view was held. Stepping outside at any given time of the day and observing the motion of the Sun, it looks like the Sun rotates around the Earth. A geocentric view of the universe was first proposed by Aristotle (384-322 BCE) and was the dominant belief held by most people for many centuries.

However, not everyone agreed with Aristotle. Aristarchus of Samos, who lived from 310-230 BCE, was an expert Greek astronomer and mathematician who did not believe that the Sun and planets revolved around the Earth. He was the first to propose a heliocentric cosmos. The word heliocentric comes from the Greek word *helios* which means "sun." A heliocentric cosmos is a view of the universe with the Sun as the central point and the Earth and planets orbiting the Sun.

Although today we know that Aristarchus was right, his proposal was rejected by his colleagues because it seemed to contradict everyday observation. If the Earth was not stable (central and not moving), how did everything not bolted down keep from flying off the Earth as it rotated around the Sun? The physics of Aristotle was the scientific consensus view during Aristarchus' lifetime and that meant that a heliocentric cosmos would have violated the laws of physics! It was almost 2000 years before the idea of a heliocentric cosmos was reintroduced by Nicolaus Copernicus (1473-1543 CE) and confirmed by the scientific observations of Galileo.

Heliocentric Cosmos

Today, astronomers do not believe in a geocentric cosmos and know that our Earth orbits the Sun and that we live in a heliocentric solar system. Modern technologies, a deeper understanding of physics, and a willingness to challenge prevailing scientific theories were needed before the geocentric view could be replaced by the more accurate heliocentric view of the cosmos.

1.5 Summary

- Astronomy is the field of science that studies celestial bodies and how they exist and move in space.

- Early astronomers were able to map the movements of the planets and stars and used celestial motions to create calendars.

- Modern astronomers use chemistry and physics together with modern technologies to study the universe.

- Ancient peoples once believed in a geocentric cosmos, or Earth-centered universe. Today we know that we live in a heliocentric solar system with the Sun at the center.

1.6 Some Things to Think About

- What do you think would make it likely that astronomy is the oldest science?

- When it gets dark, go outside and look at the stars. How many do you think you can see?

- Find a group of stars that reminds you of an object—maybe an animal. This can be your own personal constellation. What name would you give to it?

- Now that astronomers have advanced telescopes and spacecraft, do you think they have discovered everything there is to know about the stars? Why or why not?

- Why do you think it was difficult for people to accept the idea of a heliocentric cosmos?

- Do you think that as we explore space, we might find new ideas that would change the way we look at the cosmos? Why or why not?

Chapter 2   Technology in Astronomy

2.1	Introduction	9
2.2	Telescopes	10
2.3	Space Telescopes and Other Satellites	12
2.4	Other Space Tools	14
2.5	Summary	17
2.6	Some Things to Think About	18

Image credits
Background: Orion Nebula viewed by Hubble Telescope
Courtesy of NASA, ESA, T. Megeath
(University of Toledo) and M. Robberto (STScI)
Hubble Telescope Image Courtesy of NASA

Chapter 2: Technology in Astronomy   9

2.1 Introduction

In this chapter we will take a look at the tools astronomers use to explore the skies. Tools are an essential part of scientific investigation, and for many centuries astronomers have been using tools to gain a better understanding of the universe.

Even before the invention of the telescope, early astronomers used tools to study the sky. In the 1500s BCE, ancient civilizations used tools to track the movement of the Sun. Stonehenge, a group of huge stones set in a circular shape outside of Amesbury, England is believed to be a kind of solar tracking system. As the Sun moves over the large structure, the shadows mark the summer and winter solstices.

Stonehenge, near Amesbury, England

Today, modern astronomy tools help scientists get accurate readings of star and planetary movements and help them observe stellar objects that they can't see with their eyes. In this chapter we will learn about some of the tools modern astronomers use.

2.2 Telescopes

When Galileo decided to look at the night sky, he used a telescope. The word telescope comes from the Greek prefix *tele-* which means "from afar" or "far off" and *skopein* which means to "see," "watch," or "view." A telescope is an instrument used to see, watch, or view things that are far away.

Galileo is sometimes credited with the invention of the telescope, but in 1608 the Dutch lens maker Hans Lippershey filed the first patent for what would eventually become the telescope. Galileo made many improvements to the Dutch "perspective lens" and was able to greatly increase the magnification. With his powerful lenses, Galileo was able to see that Jupiter has moons!

There are essentially three types of telescopes: refractor telescopes, reflector telescopes, and compound telescopes.

The first telescopes built were refractor telescopes. Refractor telescopes can be found in hobby and toy stores and are the type of telescopes used for rifles.

A refractor telescope houses a lens at one end and an eyepiece at the opposite end of a narrow tube. Light enters one end of the tube and is bent by the lens. The observer looks through the tube from the opposite end and sees the magnified object at the focal point, the spot at which the rays of light entering the lens come together to produce the image.

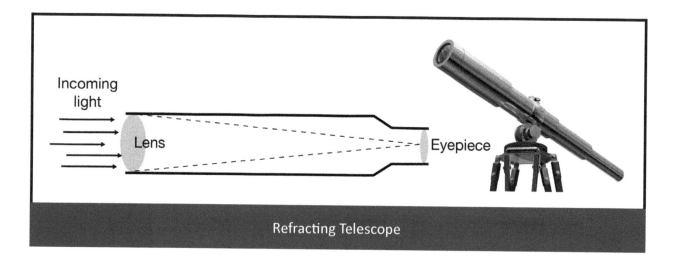

Refracting Telescope

Chapter 2: Technology in Astronomy 11

The largest refracting telescope ever constructed was at the Great Paris Exhibition in 1900. It had a focal length (the distance from a lens to its focal point) of 57 meters (187 feet) but was later dismantled after the company went bankrupt. Today the largest refracting telescope that has been recently used is housed at the Yerkes Observatory in Williams Bay, Wisconsin.

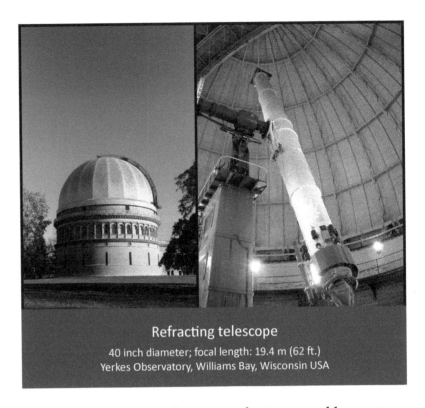

Refracting telescope
40 inch diameter; focal length: 19.4 m (62 ft.)
Yerkes Observatory, Williams Bay, Wisconsin USA

Reflector telescopes and compound telescopes use a combination of mirrors and lenses to focus incoming light. These telescopes are more complicated than the refractor telescope and can provide better quality images. A common reflector telescope is the Newtonian telescope named after its inventor, Isaac Newton. A Newtonian telescope has a simple design with two mirrors and an eyepiece. Light enters the telescope at the far end, is reflected back by one mirror, and hits a second mirror where it exits to the eyepiece.

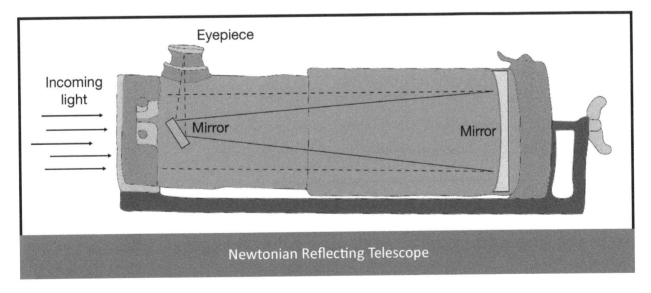

Newtonian Reflecting Telescope

A compound telescope uses elements of both reflector and refractor telescopes.

2.3 Space Telescopes and Other Satellites

Although telescopes can be built for viewing planets and stars that are millions of miles away, Earth's atmosphere causes problems for astronomers making observations from the Earth's surface. Light coming from a distant star must pass through Earth's atmosphere before it can be collected by a telescope on the surface. As the light passes through the atmosphere, it can be reflected by tiny atmospheric particles. Atmospheric turbulence causes these particles to move, creating small changes in the optical properties of the air. Before it enters a telescope, the light that is being collected gets bounced around, which makes it appear that the image of the object being viewed is moving. This also makes stars appear to "twinkle." Although twinkling stars are fun to watch, they prevent astronomers from collecting the kind of detailed data that is needed for study.

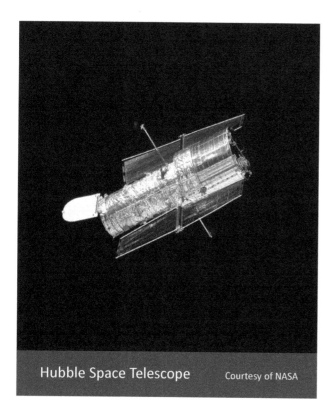

Hubble Space Telescope — Courtesy of NASA

A great way to solve this problem is to set up a telescope outside the Earth's atmosphere. The Hubble Space Telescope is one such telescope. It was placed into orbit around Earth by the Space Shuttle Discovery in 1990. The Hubble Space Telescope is able to take sharp and detailed images of many far distant objects with very little distortion. Also, space telescopes are able to view the universe in wavelengths of light that are blocked by Earth's atmosphere, such as X-rays and gamma rays, and also rays that are partially blocked by the atmosphere, such as ultraviolet and infrared rays.

Our understanding of the universe has been greatly expanded as a result of images taken by the Hubble Space Telescope. Astronomers have been able to learn more about how stars and galaxies form, old stars explode, and galaxies collide. Much has also been learned about black holes, nebulae, and planets that are orbiting other stars.

Chapter 2: Technology in Astronomy   13

Hubble Space Telescope image
A mountain of dust and gas rising in the Carina Nebula
Courtesy of NASA, ESA,
and M. Livio and the Hubble 20th Anniversary Team (STScI)

Another satellite is the Kepler Space Telescope which was placed into orbit in 2009 for the purpose of looking for planets outside our solar system. Even though Kepler explores a very small portion of the sky, it has discovered many planets, and scientists now think that most stars have planets orbiting them. We'll take another look at the Kepler Space Telescope in Chapter 8.

The largest and most complex satellite is the International Space Station (ISS). The space agencies most involved in constructing and operating the space station are from the United States, Russia, Europe, Japan, and Canada. The first space station module was launched in 1998 and more modules have been added over the years. The first crew arrived at the ISS in 2000, and it is now continuously occupied, with crews from different countries coming and going. Research done on the ISS includes studies of human health and life sciences, testing technologies that may be used in future space explorations, and research in various fields such as physical sciences and earth and space science.

A meeting of the minds aboard the International Space Station on 7 March 2015
Members of Expedition 42: astronaut US, Barry Wilmore (Commander) Top, Upside down, to the right Russian cosmonaut Elena Serova, & ESA astronaut Samantha Cristoforetti. Bottom center US astronaut Terry Virts, top left cosmonauts Alexander Samokutyaev and Anton Shkaplerov.
Courtesy of NASA

There are currently over 1000 operational satellites orbiting Earth. Satellites have become an important part of our lives, with uses for television, cell phones, GPS, weather forecasting, and monitoring changes to Earth's environment and climate. They also have many other scientific uses such as the study of the Sun, Moon, and magnetosphere. Satellites have greatly increased our knowledge of Earth and the cosmos and make it much easier for us to communicate.

2.4 Other Space Tools

Space probes, landers, and rovers are other tools scientists use to explore space.

A space probe is a robotic spaceship that can travel far distances, capturing images and collecting data. Voyager 1 and 2 are space probes that were launched in 1977 to explore Jupiter and Saturn. Voyager 1 has continued on its journey to become the first spacecraft to leave our solar system and is now traveling in interstellar space. After passing Jupiter and Saturn, Voyager 2 also went past Uranus and Neptune, sending data back to Earth, and it is now nearing interstellar space. Both Voyager 1 and 2 are still sending signals back to Earth for scientific analysis.

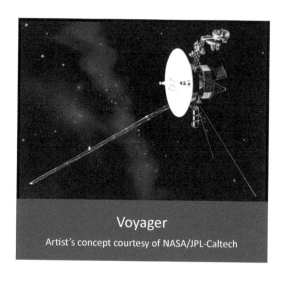
Voyager
Artist's concept courtesy of NASA/JPL-Caltech

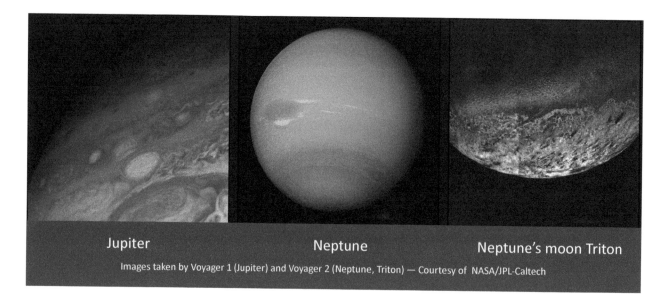
Jupiter Neptune Neptune's moon Triton
Images taken by Voyager 1 (Jupiter) and Voyager 2 (Neptune, Triton) — Courtesy of NASA/JPL-Caltech

Spacecraft that go into orbit around celestial bodies other than Earth are called orbiters. For example, NASA's Mars Atmosphere and Volatile Evolution (MAVEN, launched in 2013) is studying the atmosphere of Mars, and India's Mars Orbiter Mission (MOM, launched in 2013) is looking at the surface features and mineralogy as well as the atmosphere of Mars. Both of these orbiter missions have a fairly short life expectancy — depending on when they run out of fuel.

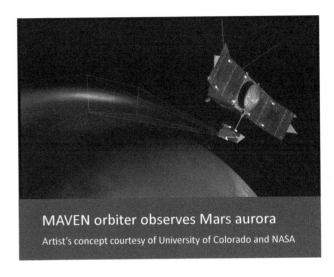

MAVEN orbiter observes Mars aurora
Artist's concept courtesy of University of Colorado and NASA

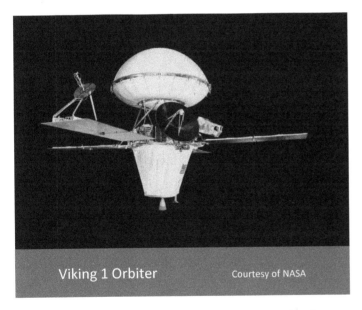

Viking 1 Orbiter   Courtesy of NASA

Like space probes and orbiters, a lander is a robotic spacecraft, but it is able to land on the surface of planets, asteroids (small celestial bodies made mostly of rock and minerals), or comets (large chunks of ice and dirt). NASA's Viking 1 was a spacecraft that consisted of both an orbiter and a lander, and the Viking 1 lander was the first to successfully land on the surface of Mars and send images back to Earth. Viking 1 touched down on Mars in July 1976 and continued collecting data for six years. The Viking 1 orbiter sent back images of Mars from space from 1976 through 1980.

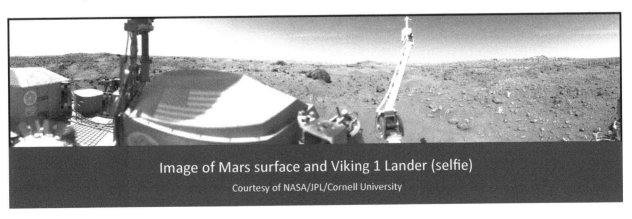

Image of Mars surface and Viking 1 Lander (selfie)
Courtesy of NASA/JPL/Cornell University

A more recent orbiter-lander combination is the Cassini-Huygens mission which is a cooperative project of NASA, the European Space Agency, and the Italian Space Agency. In 2004 the Cassini spacecraft released the Huygens lander over Saturn's largest moon, Titan. Huygens had a safe landing and sent back data that revealed many interesting and unexpected details about Titan. Cassini has gone on to explore Saturn and its moons and has also found many unexpected features, such as moons covered with ice.

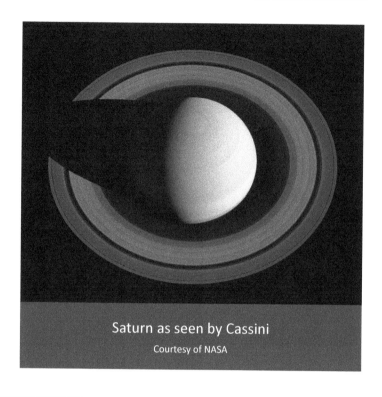

Saturn as seen by Cassini
Courtesy of NASA

Mars rover Curiosity (selfie)
Courtesy of NASA/JPL-Caltech/Malin Space Science Systems

A rover is a lander that can move. A rover is an automated machine that can travel across a planetary surface, gathering data as it goes. Since the mid 1990s several NASA rovers have landed on Mars. The first was Sojourner in 1997, followed by Spirit and Opportunity in 2004. Spirit became stuck in the sand, and in March 2010, it stopped transmitting data. In June 2018 there was a global dust storm on Mars, causing Opportunity to stop transmitting.

The most recent Mars rover is Curiosity which was launched in November 2011. Curiosity's mission includes: looking for the chemical building blocks of life, such as organic carbon compounds; investigating the chemical composition of the Martian surface; interpreting the ways rocks and soils may have formed; and looking for the cycling of carbon dioxide and water. The rovers are helping scientists learn about the makeup of Mars, which will aid in planning for astronauts to successfully land on Mars.

A hole drilled by Curiosity to collect a sample to analyze

Courtesy of NASA/JPL-Caltech/MSSS

These are just a very few examples of the many different spacecraft that have been launched by different countries. Space exploration continues to expand rapidly, with each discovery building on the last and increasing our knowledge of what lies beyond Earth.

2.5 Summary

- Astronomers use tools to explore the cosmos.
- Telescopes are used to magnify faraway objects.
- Space telescopes avoid atmospheric distortion.
- Satellites increase our knowledge of Earth and the cosmos and have become an important part of our lives.
- Modern astronomers can utilize space tools to collect data from far distant locations. Space probes, orbiters, landers, and rovers are among the tools used.

2.6 Some Things to Think About

- Do you think ancient peoples were able to track many astronomical events? Why or why not?

- Briefly describe the difference between a refracting telescope and a reflecting telescope. Which would you rather have? Why?

- How do you think space telescopes and satellites have changed life on Earth?

- What do you think scientists can learn by doing experiments on the International Space Station?

- Mars is being visited by orbiters, landers, and rovers that send data back to Earth. Do you think using different types of spacecraft makes it possible to study different features of Mars? What data do you think each type of spacecraft collects? What do you think are the advantages and disadvantages of each?

Chapter 3 Earth in Space

3.1	Introduction	20
3.2	The Earth in Space	20
3.3	The Earth and the Moon	22
3.4	The Earth and the Sun	25
3.5	Eclipses	27
3.6	Summary	28
3.7	Some Things to Think About	28

Image courtesy of NASA

3.1 Introduction

In this chapter we will start looking at celestial objects in space. The first object in space that we will explore is Earth itself.

What is the shape of the Earth? Is it flat, round, elliptical? Where does the Earth sit with respect to the larger universe? Is it in the middle, off to the side, on the outer edge? In this chapter we will explore these questions and others as we examine the Earth in space.

3.2 The Earth in Space

The ancient Greeks understood that the Earth is a ball, or spherical mass. The best evidence in ancient times for the Earth being spherical came from the observation that a circular shadow is cast during a lunar eclipse. The ancient Greeks could see the curvature of Earth from its shadow on the Moon.

However, it was only within the last 100 years that we have been able to photograph the Earth in space. The very first pictures of Earth as seen from space were taken in 1946 by a group of scientists in New Mexico. These scientists attached a 35 millimeter camera to a missile and launched the missile 65 miles into space. The missile came crashing down, but the camera was protected in a tough metal container. The crude black and white photos showed the curvature of the Earth and marked a new era of space exploration.

Earth is a planet. The word planet comes from the Greek word *planetai* which means "wanderer." A planet is a large spherical object or celestial body that "wanders" in space. To qualify as a planet, a celestial body must orbit a sun, must have enough mass to have its own gravity, and must have cleared its orbit of other celestial bodies (in other words, a planet can't have other celestial bodies with it in the same orbit around the Sun). Because Earth "wanders," or moves in space, around the Sun, is massive enough to have its own gravity, and also has cleared its orbit, Earth qualifies as a planet.

At the equator, Earth is 12,756 kilometers (7,926 miles) in diameter. Between the North Pole and the South Pole Earth's diameter is 12,714 kilometers (7,900 miles). You can see that Earth is not a perfect sphere but is slightly larger in one dimension.

Earth sits on a tilted axis, which is the imaginary line around which the Earth rotates. Having a tilted axis means that the North and South Poles are not straight up and down in relation to the Earth's orbit around the Sun. If you were to look at the planetary axis, you would see that the poles are tilted about 23 degrees from center. This deviation from perpendicular is called orbital obliquity. Orbital obliquity, the tilt of Earth's axis, gives us the seasons.

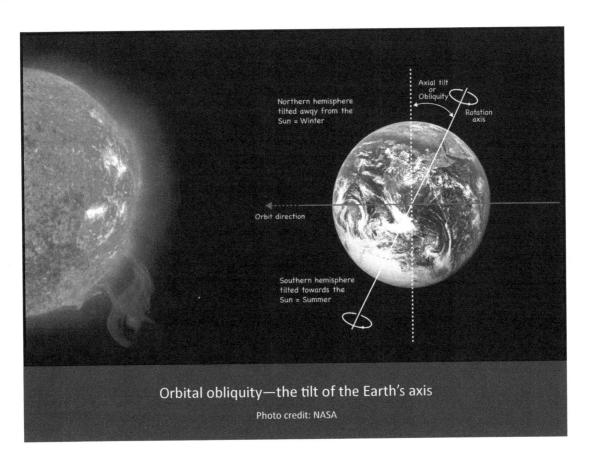

Orbital obliquity—the tilt of the Earth's axis

Photo credit: NASA

As the Earth spins, or rotates, on its axis, different parts of the Earth face toward or away from the Sun. The Earth actually makes one full rotation around its axis in slightly less than 24 hours, at 23.93 hours. This rotation on a roughly 24 hour cycle gives us our days and nights.

During different seasons of the year, the North and South Poles get more sunlight or less sunlight than the areas around the equator because the tilt of the axis points a pole toward or away from the Sun. Because of this, the poles can have nearly 24 hours of sunlight or 24 hours of darkness. So, not all days and nights are equal everywhere on the planet.

3.3 The Earth and the Moon

The Earth has one moon. A moon is any celestial body that orbits a planet. A moon is also called a natural satellite. The word moon comes from the Greek word *menas* which means month. The Moon orbits the Earth and completes one orbit every 27 days (roughly one month), hence its name — the "Moon."

The Moon can be seen from Earth because the Moon reflects light from the Sun. As the Moon orbits the Earth and as both the Earth and the Moon orbit the Sun, the appearance of the Moon changes. We call these changes phases.

In the first phase, on Day 0, the Moon is called a new moon. The new moon occurs when the Moon is between the Earth and the Sun. Only the back side of the Moon is illuminated by the Sun, so from Earth the Moon looks dark.

Phases of the Moon
Image credit: NASA/nasaimages.org

As the Moon continues to orbit the Earth, by Day 4 it enters the next phase called the waxing crescent moon. From Earth, only a small portion of the Moon appears illuminated and is crescent shaped. By Day 7 the Moon moves to the next phase and appears half-illuminated, This is called the first quarter moon, or half-moon. A few days later, on Day 10, the Moon moves to the waxing gibbous phase. A gibbous moon is between a full moon and a quarter moon. The word gibbous means "marked by convexity or swelling," so a gibbous moon is a moon that looks "swollen."

By Day 14, the Moon enters the full moon phase. The Moon is now on the opposite side of the Earth from the Sun and is seen with full illumination. A few days later, by Day 18, the Moon becomes a waning gibbous moon. Then, by Day 22, it enters the next to last phase, the last quarter moon, when the Moon is again half-illuminated, but now the illumination appears on the opposite side from the first quarter moon. Finally, by Day 26 the Moon enters the last phase, becoming a waning crescent moon. By Day 30 the Moon is back to being a new moon and the cycle repeats.

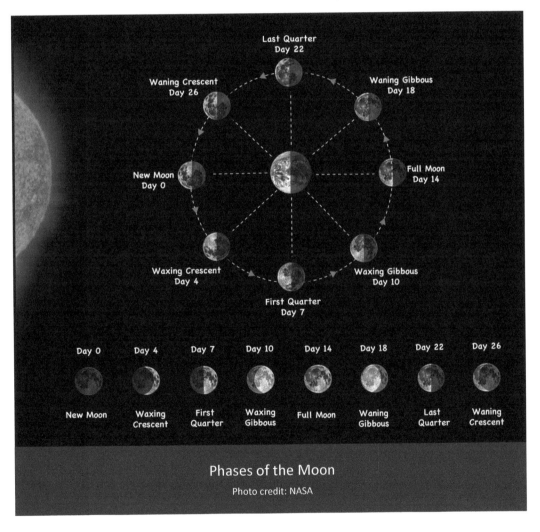

Phases of the Moon
Photo credit: NASA

The Moon and the Earth interact with each other through long range gravitational forces. Any object that has mass (the property that makes matter resist being moved) also has gravitational force (the force exerted by objects on one another). Your body has mass and also a small amount of gravitational force. But because you are very small compared to the Earth, your gravitational force does not affect the Earth.

The Moon is much bigger than you. But it has much less mass than the Earth and therefore has less gravitational force. However, the Moon has enough mass to create a gravitational pull on the Earth.

The Moon has dramatic effects on Earth. For example, the Moon is believed to stabilize Earth's rotation and the tilt of its axis. Without a moon, the Earth might swing more dramatically between degrees of obliquity, unable to maintain an average tilt of 23 degrees. If the Earth tilted more or less dramatically, this could result in extreme or even catastrophic changes in the seasons.

The Moon also contributes to the rise and fall of ocean tides. Ocean tides on Earth are created in part by the gravitational forces exerted by the Moon. The Moon (together with the Sun) pulls on the Earth's center, which creates two tidal bulges. As the Earth rotates on its own axis, these bulges are dragged along the Earth's surface, causing the sea level to rise and fall, thus creating tides.

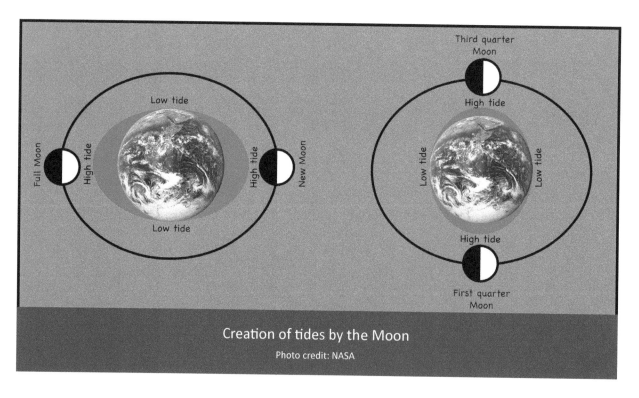

Creation of tides by the Moon
Photo credit: NASA

3.4 The Earth and the Sun

The Sun is a celestial body in space. It provides the Earth with power. The Sun is like a big battery that never runs out, continuously giving us energy in the form of light and heat. From this energy, life is possible. Without the Sun there would be no plants, animals, reptiles, fish, or even microbes. All of life requires energy in order to move, grow, eat, and reproduce. Every chemical reaction in your body requires energy, and it is ultimately the Sun's energy that powers the chemical reactions in your body.

Not only does the Sun power our planet, it also interacts with Earth, affecting tides, weather, and even our magnetic field. We saw in the last section how the Moon pulls on the Earth causing tides in our oceans. The Sun also pulls on the Earth causing tidal activity.

Did you know that "space weather" affects our own weather on Earth? It's easy to forget that Earth is not a closed system. We are a blue ball in space, interacting with other space objects like planets and the Sun. The Sun affects our planet in major ways, and one way is the weather.

Weather can be tough to predict on Earth. You might not think that a solar storm on the Sun could cause a storm on our planet, yet this is exactly what happens. Earth's weather is caused by temperature and moisture variations in different places. When the Sun has a solar storm and a burst of heat escapes, we get a rise in temperature on Earth, which can then create storms.

Solar storm
Credit: NASA/nasaimages.org

The Sun also interacts with Earth's atmosphere causing auroras, which are sometimes called northern lights and southern lights. Auroras are caused by solar storms that charge particles in space. These charged particles get trapped by Earth's magnetic field. When this happens, they pass through our atmosphere and give off light as they release energy.

Some different shapes of aurorae boreales seen from Iceland at the Arctic Circle

Photo Credit: Schnuffel2002, CC BY SA 4.0, Wikimedia Commons

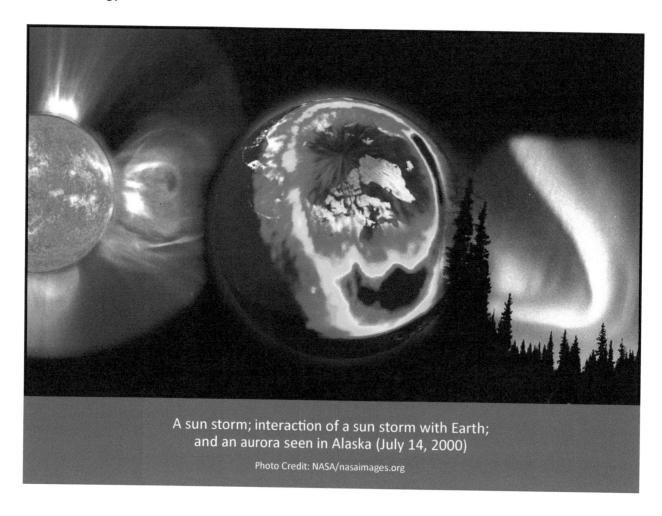

A sun storm; interaction of a sun storm with Earth; and an aurora seen in Alaska (July 14, 2000)

Photo Credit: NASA/nasaimages.org

3.5 Eclipses

There are two types of eclipses that occur. A lunar eclipse occurs when the Moon passes directly behind the Earth and the Earth blocks the Sun's rays from illuminating the Moon. The Moon is darkened as the Sun's rays are blocked and the Earth's shadow passes across the Moon.

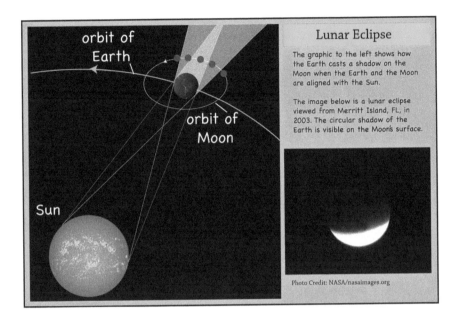

Lunar Eclipse

The graphic to the left shows how the Earth casts a shadow on the Moon when the Earth and the Moon are aligned with the Sun.

The image below is a lunar eclipse viewed from Merritt Island, FL, in 2003. The circular shadow of the Earth is visible on the Moon's surface.

Photo Credit: NASA/nasaimages.org

The other type of eclipse is called a solar eclipse. A solar eclipse occurs when the Moon passes between the Sun and the Earth, blocking the Sun's rays from reaching some location on Earth.

It is tempting to look at a solar eclipse with your naked eye. However, it is very dangerous to look at the eclipse directly. Special glasses or projection techniques must always be used to view a solar eclipse.

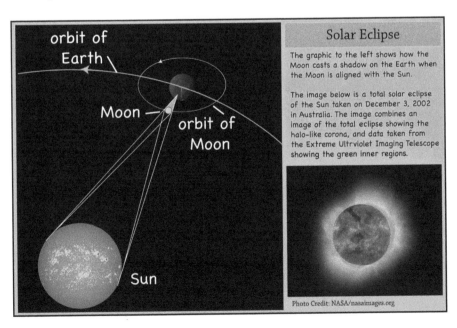

Solar Eclipse

The graphic to the left shows how the Moon casts a shadow on the Earth when the Moon is aligned with the Sun.

The image below is a total solar eclipse of the Sun taken on December 3, 2002 in Australia. The image combines an image of the total eclipse showing the halo-like corona, and data taken from the Extreme Ultrviolet Imaging Telescope showing the green inner regions.

Photo Credit: NASA/nasaimages.org

3.6 Summary

- Earth is classified as a planet because it rotates around the Sun, has enough mass to have its own gravity, and has cleared its orbit.

- Earth rotates on a tilted axis. This tilt is called orbital obliquity. The rotation of Earth on its axis gives us night and day, and orbital obliquity creates the different seasons.

- The Earth has one moon orbiting it. The Moon stabilizes the tilt and rotation of Earth and contributes to the activity of the tides.

- The Earth orbits the Sun. The Sun provides Earth with energy and contributes to Earth's weather and tidal activity.

3.7 Some Things to Think About

- How do you think Earth fits into the rest of the universe?

- What do you think it would be like to live in your town if Earth was not tilted? What would it be like at the North and South Poles?

- In your own words, explain why the Moon appears to change shape over the course of a month.

- What do you think would happen if Earth was closer to the Sun? If it was farther away? Why?

- If the Moon's orbit was tilted with respect to the Sun, would eclipses still occur? Why or why not?

Chapter 4 The Moon and the Sun

4.1	Introduction	30
4.2	The Moon	30
4.3	The Sun	33
4.4	Chemistry and Physics of Stars	34
4.5	Summary	36
4.6	Some Things to Think About	36

4.1 Introduction

As we saw in Chapter 3, the Moon and the Sun play an important role in many of Earth's properties, including the Earth's stability, rotation, weather, and tidal actions. In this chapter, as we move away from the Earth and start exploring space, we will take a closer look at the Moon and the Sun.

4.2 The Moon

Recall that the word "moon" comes from the Greek word *menas* which means month. We call our moon "the Moon" because it orbits the Earth in a monthly cycle, but not all moons orbit their planets in a monthly cycle.

The adjective lunar is also used to refer to the Moon. Lunar comes from the Latin word *luceo* which means "to shine bright." Although the Moon does not create its own light, it reflects the Sun's light and is the brightest object in the evening sky.

The Moon   Photo Credit: NASA/nasaimages.org

The Moon is not made of green cheese. In fact, the Moon is made of elements and minerals, just like Earth. We know that the surface of the Moon is made of elements and minerals because Moon samples were collected by the Apollo astronauts between 1969 and 1972. Moon soil contains aluminum, calcium, iron, magnesium, silicon, and titanium.

There are at least two different types of Moon rocks, basaltic rocks and breccia. Basaltic rocks were formed by the hardening of lava that came from lunar volcanoes and from lava that flowed through cracks that occurred in the Moon's surface when meteorites struck it. A meteorite is a rocky object in space that has hit the surface of another object in space. Rocks called breccia were formed from soil and pieces of rock that were squeezed and melted by the extreme pressure created when objects such as meteorites hit the Moon.

The Moon has little or no atmosphere, and in fact, the space around the Moon is close to being a vacuum. A vacuum is an area that contains no matter. The result of the lack of atmosphere is that there is no weather on the Moon. There are no clouds, rain, or wind. However, water in the form of ice has been discovered below the surface of the Moon.

The Moon is about 3.5 times smaller than the Earth, with a diameter of 3475 kilometers (2158 miles). However, in our solar system it is relatively the largest moon compared to the size of the planet orbited.

Because the Moon has less mass than Earth, it also has less gravity. The gravitational force on the Moon is about one sixth the gravitational force on Earth.

It takes the Moon the same number of days to complete one rotation on its axis as it takes the Moon to make one orbit around the Earth. This means that as the Moon is going around the Earth, the side of the Moon that faces the Earth is slowly rotating at exactly the same rate that the Moon is orbiting the Earth. Therefore, the Moon always has the same side facing Earth.

The Earth and the Moon
Photo Credit: NASA

The temperature on the Moon varies wildly because it has no atmosphere to hold on to the heat from the Sun. During the day, temperatures can be as high as 173 degrees Celsius (280 degrees Fahrenheit). At night, the temperature can dip to as low as -240 degrees C (-400 degrees F).

If you look up at the Moon, you will see both light and dark areas on the surface. The light areas are known as terrae. The word *terrae* is Latin and means "lands." These light areas are rugged with craters that can exceed 40 kilometers (25 miles) in diameter.

The dark areas on the Moon are known as maria. The word maria comes from the Latin word *marinus* which means "sea." When 17th century astronomers were looking at the Moon through their telescopes, they thought that the large dark areas were bodies of water, or seas.

The early astronomers gave the maria fun names such as Mare Tranquillitatis, meaning "The Sea of Tranquility," and Mare Nectaris, meaning "The Sea of Nectar." Modern astronomers still use these names but know that the dark areas of the Moon are not seas. Instead, these dark areas are lava flows.

Like Earth, the Moon is thought to be made of a crust (an outer, rocky shell), a mantle (the layer below the crust), and an iron-rich core. However, the Moon now has no magnetic field. By studying rocks brought back from the Moon by the Apollo astronauts, scientists have discovered that the Moon had a strong magnetic field in the far distant past. Scientists don't know for sure why the Moon lost its magnetic field, but they are exploring different theories to discover the cause.

The Moon—crust and interior
Photo credit: NASA

The Apollo astronauts placed seismometers on the Moon's surface. Seismometers are instruments that detect motion or vibration in the ground. By using these seismometers, scientists found that the Moon is seismically active and can have earthquakes that last longer than 10 minutes — much longer than those on Earth. It is thought that the Moon does not have moving tectonic plates like Earth but that earthquakes may be caused by events such as meteorites striking the Moon's surface and the sides of craters collapsing, among other things.

4.3 The Sun

As we saw in Chapter 3, the Sun affects Earth's tides, weather, stability and rotation. But what is the Sun?

The Sun is different from both the Earth and Moon. The sun is a star. A star is a celestial body that generates light and heat energy. Our star, the Sun, is composed mainly of hydrogen and helium. Hydrogen and helium are the lightest elements known and are gases. Hence, one way to think about the Sun is to imagine it as a hot ball of gas.

Although the Sun is made of lightweight gases, the Sun's diameter is about 100 times the diameter of the Earth, and the Sun is over three hundred thousand times as massive as the Earth.

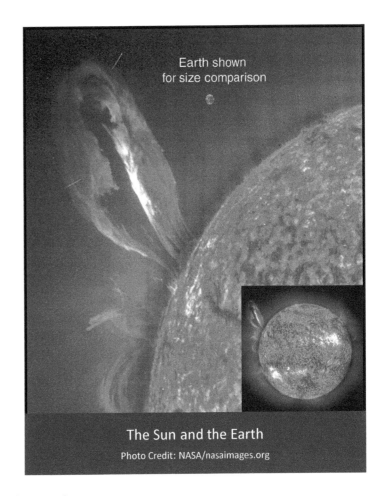

The Sun and the Earth
Photo Credit: NASA/nasaimages.org

Mass of Sun/Mass of Earth = 332,840

To get an idea of just how large the Sun is compared to the Earth, know that about 1 million Earths would fit inside the Sun!

The Sun's temperature is extremely hot, averaging 5800 degrees kelvin (5510° Celsius or 9900° Fahrenheit) with some regions exceeding tens of thousands or even millions of degrees kelvin. Life on Earth is made possible by the extreme temperatures on the Sun that radiate into space. But how does the Sun generate so much energy?

4.4 Chemistry and Physics of Stars

To understand how the Sun can provide the energy to fuel our planet, it is important to look at the chemistry and physics of stars.

As we saw in Section 4.3, the Sun is composed largely of hydrogen and helium, two lightweight gases. When gases are compressed (squeezed into a smaller space), their temperature increases. If you hold your hand on a tire tube as you pump air into it, you will find that the tube gets warm. This heat is generated by the increasing pressure imposed on the gas (air) as more gas molecules are forced into the tube.

The Sun is a huge ball of compressed gases and has extremely high temperatures at its center. It is believed that these temperatures are so high that the hydrogen atoms become ionized. Ionization is a process where the electrons and nucleus of an atom become separated. Since a hydrogen atom is made of one proton, one electron, and no neutrons, when hydrogen ionizes, it is converted to a free proton (the nucleus) and a free electron. The proton is positively charged and the electron is negatively charged.

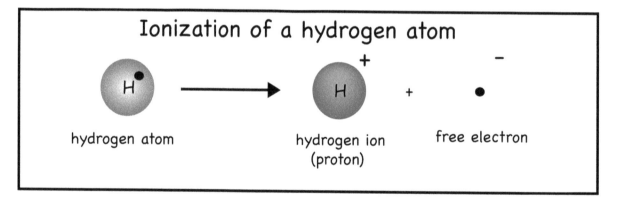

What happens when free hydrogen protons meet? We know that when two like charges meet they will repel each other. However, at the extreme pressure and temperature that exist at the center of the Sun, when two hydrogen protons meet, they combine, or fuse, together. This process is called hydrogen fusion. The fusing together of two protons is also called thermonuclear fusion because it can take place only at extremely high temperatures.

However, not only do the two hydrogen protons fuse, but one of the two is converted into a neutron. Eventually, four hydrogen protons (nuclei) will fuse to make a single helium atom.

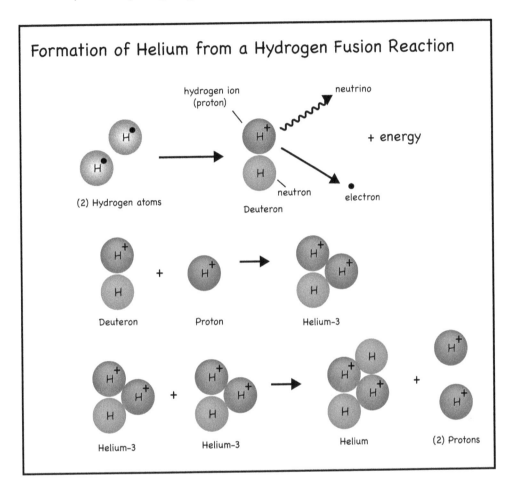

Thermonuclear fusion is a nuclear reaction that releases massive amounts of energy. In 1905 Albert Einstein showed that mass can be converted to energy with his elegantly simple equation:

$$E = mc^2$$

The symbol "E" represents "energy," the symbol "m" represents "mass," and the symbol "c" represents "the speed of light." Because the speed of light is a very large number (c = 299,792,458 x meters/seconds) and is multiplied times itself ("c^2") and then multiplied times the mass ("m"), only a small amount of mass is needed to create large amounts of energy.

For example, if you had a mass ("m") of one gram of hydrogen, using the formula $E=mc^2$ you would get a result of 21,480,248,771,809 calories of energy ("E"), with calorie being the unit of measurement of energy. This means that one small gram of hydrogen would give 21 trillion units of energy!

A chocolate chip cookie contains 227 calories. If you eat a chocolate chip cookie, your body gets 227 units of energy. Your body needs energy to do things like ride a bike or row a boat. To get 21 trillion units of energy you would need to eat about 94 billion chocolate chip cookies!

Thermonuclear fusion is the process the Sun uses to convert hydrogen into helium and energy. With the tremendous amounts of energy thermonuclear fusion creates, the Sun can fuel our entire planet!

4.5 Summary

- The Moon is made of the same elements found on Earth.
- The Moon is smaller than the Earth and has little or no atmosphere and no liquid water.
- The light areas of the Moon are called terrae and the dark areas are called maria.
- The Sun is a large celestial body composed mainly of the two gases, hydrogen and helium.
- The Sun converts hydrogen to helium and generates energy using thermonuclear fusion.

4.6 Some Things to Think About

- Do you think life could exist on Earth without the Moon?
- Review the descriptions of the Moon. If you wanted to live on the Moon for a while, what problems would you have to solve to create a place where you could survive?
- What are some differences between the Sun and the Moon?
- Do you think the Sun could ever run out of energy? Why or why not?

Chapter 5 Planets

5.1	Introduction	38
5.2	Types of Planets	38
5.3	Earth-like Planets	39
5.4	Jupiter-like Planets	41
5.5	What Happened to Pluto?	44
5.6	Summary	45
5.7	Some Things to Think About	46

Images courtesy of NASA

5.1 Introduction

Up to this point we have been introduced to the Earth, the Moon, and the Sun. Although all of these celestial bodies are made of the same elements as those found on Earth, they also differ from each other in significant ways. We discovered that the Earth is a planet, the Moon is a moon, and the Sun is a star. In this chapter we will take a look at different types of planets.

5.2 Types of Planets

Each of the planets that orbits the Sun is unique. Earth is the only planet that has liquid water and an atmosphere that is breathable by humans. Jupiter is the only planet where immense storm systems last for centuries, and Venus has a cloud layer made of sulfuric acid.

In our solar system there are officially eight planets (Pluto lost its status as a planet, and we'll find out why in Section 5.5). The names of the eight planets that orbit our Sun are Mercury, Venus, Earth, Mars, Jupiter, Saturn, Uranus, and Neptune.

Although the planets differ greatly from one another, they can be placed into two broad categories: the terrestrial planets (Earth-like) and the Jovian planets (Jupiter-like).

The Planets
Image Credit: NASA/nasaimages.org

The terrestrial planets differ from the Jovian planets in their physical properties and distance from the Sun. The terrestrial planets are all made of rocky materials and are relatively close to the Sun compared to the Jovian planets. The Jovian planets are made primarily of helium and hydrogen and are at greater distances from the Sun.

5.3 Earth-like Planets

The terrestrial planets include Mercury, Venus, Earth, and Mars. The word terrestrial comes from the Latin word *terra* which means "earth."

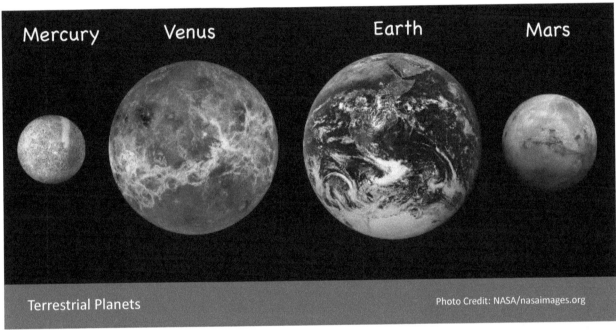

Terrestrial Planets

Photo Credit: NASA/nasaimages.org

Mercury

Photo Credit: NASA/nasaimages.org

All of the terrestrial planets resemble Earth in some ways. The terrestrial planets have hard, rocky surfaces with mountains, craters, and volcanoes.

Mercury is the terrestrial planet closest to the Sun. Because it is so close to the Sun, it is difficult to get images of it from Earth. However, in 1974 and 1975 an unmanned spacecraft called Mariner 10 got close enough to Mercury to take photographs. The Mariner 10 space probe revealed Mercury's rough, cratered surface. Although Mercury looks similar to the Moon, Mercury does not have the light and dark areas (terrae and maria) seen on the surface of the Moon.

Venus is the next closest planet to orbit the Sun. Venus looks deceptively like Earth in size and shape. For years scientists thought Venus might be a warm jungle version of Earth with teeming life. But modern technologies have given us more information about Venus, and today we know that Venus is inhospitable for life as we know it.

The atmosphere on Venus is composed almost entirely of carbon dioxide. The high level of carbon dioxide creates a surface temperature of 460 degrees Celsius (860 degrees Fahrenheit). The clouds that cover Venus contain corrosive sulfuric acid.

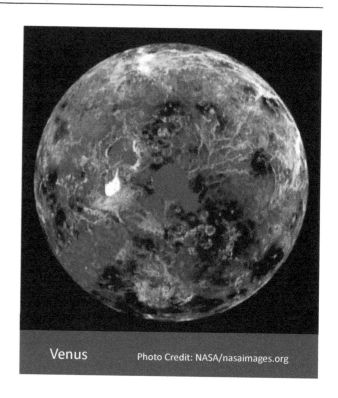
Venus Photo Credit: NASA/nasaimages.org

Earth is the third terrestrial planet from the Sun.

The fourth terrestrial planet from the Sun is Mars. People have been fascinated by Mars and have long speculated that life on Mars exists. "Martians" are a favorite character in many science fiction novels and films. Early astronomers even suggested that features on Mars included linear canals, denoting intelligent life and liquid water. Today we have found that Mars has water in the form of ice, and some scientists think they may find bacteria on Mars, but there are no advanced forms of life.

Mars is about half the diameter of Earth [6794 km (4220 mi)] with a relatively thin, almost cloudless atmosphere. Mars appears bright red to Earth observers, but the surface of Mars is actually reddish-brown. In the late 1960s several unmanned spacecraft flew past the surface of Mars

Mars Photo Credit: NASA/nasaimages.org

and sent back the first close-up pictures of the Martian surface, showing that it is covered with craters. More recently, rovers have landed on Mars and sent back to Earth photos and much data about Mars.

5.4 Jupiter-like Planets

The Jupiter-like, or Jovian, planets include Jupiter, Saturn, Uranus, and Neptune. The term Jovian comes from Roman mythological stories about Jove, who was the god of the sky. The Jovian planets are those planets that resemble Jupiter in their physical properties and distance from the Sun.

Jovian planets

Photo Credit: NASA/nasaimages.org

Jupiter is the largest of the Jovian planets, with a diameter about 11 times larger than that of the Earth. It is also about 318 times more massive. Jupiter orbits the Sun very slowly, taking almost 12 Earth years to make one orbit.

Looking at Jupiter through an Earth-based telescope, you can see light and dark bands circling the planetary surface. The light bands are called zones, and the dark bands are called belts. The zones and belts are parallel to Jupiter's equator and are colored red, orange, yellow, and brown. The zones and belts are gases at various temperatures. Scientists believe that the zones appear lighter because the clouds are higher and colder in this region, and the belts appear darker because the clouds are lower and warmer.

The Great Red Spot appears to be a huge long-lived storm where clouds complete a counterclockwise rotation about every six days. Jupiter is composed primarily of hydrogen and helium and has no rocky surface to break up the storm which has lasted for centuries.

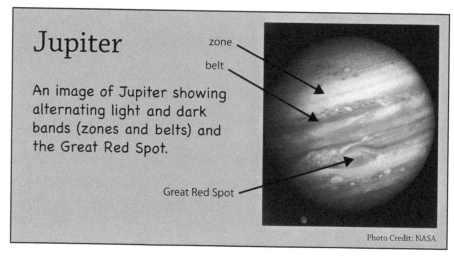

Jupiter

An image of Jupiter showing alternating light and dark bands (zones and belts) and the Great Red Spot.

Photo Credit: NASA

Saturn is the next largest Jovian planet. With a diameter nearly 9 times that of Earth, it is about 95 times more massive than Earth. Saturn, like Jupiter, is a mostly gaseous planet that slowly orbits the Sun, taking 29 Earth years to make one orbit.

Saturn

Photo Credit: NASA/nasaimages.org

Like Jupiter, Saturn has belts and zones resulting from different gas clouds at different heights and temperatures. In addition to the belts and zones, Saturn has many colored rings extending laterally from the equator. Saturn's rings are believed to be made of many millions of icy fragments that are not connected but uniformly circle the planet. These icy fragments reflect the Sun's light, causing them to illuminate brightly.

The last of the Jovian planets are Uranus and Neptune. Uranus and Neptune extend to the darkest edge of our solar system and are many millions of miles away from the Sun.

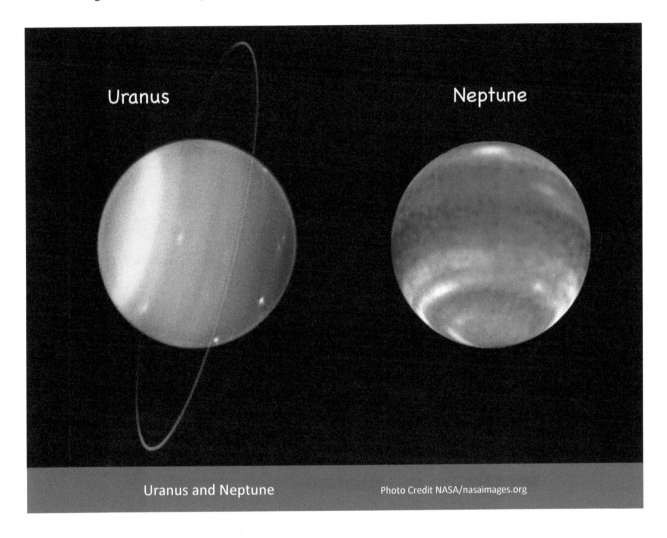
Uranus and Neptune Photo Credit NASA/nasaimages.org

Uranus rotates around the Sun every 84 Earth years, is just over 4 times the diameter of Earth, and is 14 times as massive. Uranus has an almost featureless surface without zones and belts. Uranus is made up of mostly hydrogen and helium, like Jupiter and Saturn, but also contains a significant amount of methane. The presence of methane gives Uranus a distinctive bluish color. Computer imaging shows some very faint banding which is thought to be the result of sunlight breaking down the methane gas on the planet's surface. Uranus sits on its side as it orbits the Sun and, like Saturn, it is circled by several rings.

Uranus is colder than either Jupiter or Saturn. Because of its low temperature, Uranus does not have dense clouds like those of Jupiter and Saturn, and this may explain its very bland, featureless surface.

Neptune rotates around the Sun every 164 Earth years and like Uranus is about 4 times the diameter of Earth. But Neptune differs from Uranus in that it is 17 times more massive than Earth and has Jupiter-like atmospheric clouds, belts, and zones. In 1989 the unmanned spacecraft Voyager 2 flew past Neptune and captured images of a giant storm called the Great Dark Spot. The Great Dark Spot was similar in many ways to Jupiter's Great Red Spot but was not as long-lived. In 1994 telescope images revealed that the storm had disappeared.

5.5 What Happened to Pluto?

Until recently Pluto was a favorite planet for many celestial enthusiasts, both young and old. Until August 2006 Pluto, at the far outer edge of our solar system, was considered the tiniest of the planets.

Pluto—size comparison (diam.)
Image Credit: NASA/JPL-Caltech/R. Hurt (SSC)

Pluto was discovered by Clyde Tombaugh on January 23, 1930. Tombaugh was curious about what appeared to be irregularities in the orbits of both Uranus and Neptune. Many astronomers of Tombaugh's day were troubled by the fact that Uranus and Neptune did not orbit the Sun as the astronomers thought they should. Tombaugh, knowing that neighboring planets can disturb planetary motions, found Pluto as a dim speck among the stars. Pluto was immediately given the title of the 9th planet in our solar system.

But, as it turns out, the naming of Pluto as a planet was a mistake. It was later determined that Pluto is too small to disturb the orbits of Uranus and Neptune. Also, astronomers decided there was nothing wrong with their orbits in the first place.

In 2006 the International Astronomers Union (IAU) had a meeting to discuss the definition of a planet. At this meeting it was decided that in order to qualify as a planet, a celestial body must have "cleared the neighborhood around its orbit." In other words, it can't have other celestial bodies orbiting the Sun with it. Pluto is actually in a belt of other celestial bodies called the Kuiper Belt. So the IAU reclassified Pluto as a dwarf planet rather than a true planet. A dwarf planet has not cleared its orbit but does have enough gravity to have formed a spherical shape like a true planet.

However, as is typical in science, the debate continues. As of the writing of this text, the IAU has come up with a new classification for Pluto called a plutoid. A plutoid is like a dwarf planet, but its orbit is beyond that of Neptune. Other scientists don't accept the IAU's definitions and would like to have Pluto reinstated as a planet. Who knows — a future young astronomer may help Pluto regain its planetary status!

Solar system (as of 2006) with eight planets
Image credit: International Astronomical Union (www.iau.org)

5.6 Summary

- There are officially eight planets in our solar system: Mercury, Venus, Earth, Mars, Jupiter, Saturn, Uranus, and Neptune.

- The terrestrial planets are "earth-like" (made up of mostly rock) and are Mercury, Venus, Earth, and Mars.

- The Jovian planets are "Jupiter-like" (made up of mostly hydrogen and helium) and are Jupiter, Saturn, Uranus, and Neptune.

- Pluto was considered the 9th planet in the solar system, but lost its planetary status in 2006. It is now considered a dwarf planet or a plutoid.

5.7 Some Things to Think About

- Do you think all the planets are the same? Are they more like the Moon or more like the Sun? Are they something different?

- Do you think humans might someday live on one of the terrestrial planets? On one of the Jovian planets? Why or why not?

- Some groups of people are trying to figure out how to start a colony on Mars. What problems do you think they would have to solve to be able to do this?

- In what ways are the Jovian planets like each other? In what ways are they different from the terrestrial planets?

- Why do you think there is debate about whether Pluto is or is not a planet?

- Why do you think the term plutoid was invented?

- Do you think it is important to have an organization of scientists like the IAU that makes rules and decisions about how astronomical objects are classified? Why or why not?

Chapter 6 Time, Clocks, and the Stars

6.1	Introduction	48
6.2	Reading a Star Atlas	50
6.3	Time	52
6.4	Celestial Clocks	55
6.5	Summary	57
6.6	Some Things to Think About	57

6.1 Introduction

What would you do if you were on a hike with a group of friends and suddenly found yourself separated? Imagine that your compass needle is stuck and as you take your whistle from your pocket to call your friends, it falls out of sight in a crack between two big rocks that are too heavy to lift or move. The day is coming to a close and the night stars are beginning to shine. If you don't know how to navigate with the stars, the best thing to do would be to sit down and hope that your friends or someone else will find you. However, if you do know about stars, the constellations, and how to interpret their position in the night sky, you could find your way back to the group camp or find your way home.

In ancient times people looked to the stars for inspiration, religious meaning, and navigation. One way early people gave meaning to the stars was to look for patterns, create groups of stars, and give them names. These named groups of stars are called constellations.

We don't know the exact date the first constellations were named, but ancient Egyptians, Sumerians, and Chaldeans are believed to have known many of our present-day constellations. It appears that by 2000 BCE most of the main constellations in the Northern Hemisphere had been recorded. The Greeks and Romans took over where their ancient ancestors left off, using Greek and Latin to name many of the constellations after heroes, animals, and mythical objects.

Mapping the stars is called celestial cartography or uranography. Cartography is the art and science of making maps. Uranography comes from the Greek word *uranos* which means "heavens" and *graphe*, which means "to write," so uranography means the "writing of the heavens." Although constellations are visible from both the Northern and Southern Hemispheres, it wasn't until after the 15th century CE that constellations in the Southern Hemisphere were recorded by European explorers.

There are forty-eight original constellations which include the constellations known to ancient Greek, Roman, and western Asian people. Today, the IAU (International Astronomical Union) recognizes 88 constellations. The newer constellations are found mostly in the Southern Hemisphere and were charted by Europeans as they explored that part of the globe. However, we now know that ancient cultures all over the world recognized constellations.

Mapping the stars was a popular activity for many early astronomers, and with the invention of the telescope, modern astronomers began to focus primarily on determining the accurate position of stars and celestial objects rather than using constellations for reference. In 1875 the German astronomer Friedrich Argelander (1799-1875 CE) published a catalog of the locations of 325,000 stars. This star catalog, called the Bonner Durchmusterung, was simply a grid that showed the positions and magnitudes (brightness) of stars without relating them to the constellations. This star catalog is still being used in revised and updated forms. Another star catalog still in use today is Norton's Star Atlas which was first published in 1910 and was based on a star catalog created by Belgian astronomer Jean-Charles Houzeau (1820-1888 CE). Norton's Star Atlas has also been revised and updated many times.

From 1989-1993 the Hipparcos satellite gathered data that was used to create the Hipparcos Star Catalog, which accurately mapped over 100,000 of the brightest stars, and the Tycho Star Catalog, which mapped over 2 million dimmer stars with slightly less accuracy. The Hubble Space Telescope has also been used to catalog stars.

Today, there are different star catalogs, or star atlases. In addition to the Hubble and Hipparcos/Tycho star catalogs, there are a number of print and computer generated star atlases that map not only the stars in our galaxy, but deep space stars, nebulae, and celestial objects. Many star maps include the constellations in addition to the positions, magnitudes, and movement of individual stars.

6.2 Reading a Star Atlas

A star atlas or star map can be quite overwhelming for the beginning astronomer. Modern star atlases map over 450,000 stars, planets, and other celestial objects. Just locating where you are relative to the stars in the sky can be daunting. Not only are there thousands of tiny dots representing the locations and brightness of the stars, but the stars' locations in the sky are constantly changing with the seasons.

For this reason, it's easiest to orient yourself starting with familiar landmarks like the constellations and asterisms, which are groups of stars that are smaller than constellations and may be part of a constellation. Patterns are easier to see than a single star in a cluster of stars, and by locating a constellation or asterism in the sky you can find the surrounding stars mapped in the atlas. For example, if you are in the Northern Hemisphere, one of the easiest asterisms to locate is the Big Dipper. Recall that the two stars in the part of the bowl of the Big Dipper that are farthest from the handle point to Polaris, the North Star. When you are facing Polaris, you are facing north. Polaris is in between the Big Dipper and the constellation called Cassiopeia. Cassiopeia is a W-shaped set of stars that is easy to find.

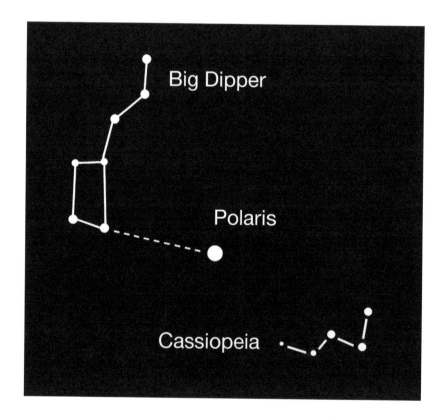

Chapter 6: Time, Clocks, and the Stars   51

Once you find the Big Dipper, Polaris, and Cassiopeia, you can use the star atlas as a kind of road map. By holding the star map above your head, and turning it to align with the constellation landmarks, you will be able to identify stars that are outside the constellation. Star maps vary with a given day or season since the stars change position throughout the year and some are only visible at certain times of the year.

Star atlases also come as calendar charts. Each calendar chart lists the stars and constellations that are visible from different locations on Earth and shows the positions of the stars for a particular month. Some star calendars show thousands of stars, but many star calendars only show a few hundred stars, making it easier to navigate.

Star atlases not only map the location of the stars but indicate a star's brightness. Brighter stars are shown as large dots, with less bright stars shown as smaller dots. Star atlases also map galaxies, clusters, and nebulae using different symbols such as dotted circles, closed circles, and ovals.

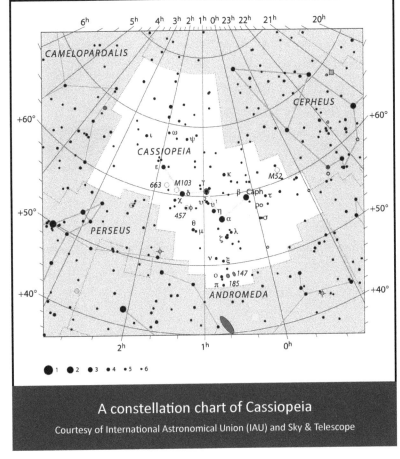

A constellation chart of Cassiopeia
Courtesy of International Astronomical Union (IAU) and Sky & Telescope

6.3 Time

What time is it? How do you determine the time when you need to get up in the morning, go to class during the day, or go to sleep at night? If you are like most people in the modern world, you probably use your wristwatch, wall clock, or digital time on your phone. But what is time exactly and how do you know your watch is correct?

In the morning you can see the Sun rise and in the evening you can see the Sun set. The Sun comes up over the eastern horizon and sets in the western horizon after spending a certain amount of time in the sky illuminating your day. The time it takes for the Sun to go from its highest position in the sky on one day to its highest position on the next day is called a solar day. If you use a sundial, you can measure when the Sun is at its highest position in the sky and how long it takes the Sun to go from one position in the sky to the next. A sundial records the Sun's daily motion across the sky and gives apparent solar time. Apparent solar time is time measured from the direct observation of the Sun. But on cloudy days and at night a sundial won't work. Although apparent solar time is a natural way to keep track of time, it isn't accurate enough for our modern world, so other ways of keeping time have been devised.

An example of a sundial

If you live anywhere on Earth other than near the equator, you will notice that due to the tilt of Earth on its axis the length of a day changes with the seasons. In the Northern and Southern Hemispheres during the winter months the days are shorter than in the summer months. At the equator the length of a day stays about the same all year, and the farther you are from the equator, the more variation there is in the length of the day throughout the seasons. Also, because Earth's orbit is slightly elliptical, Earth's distance from the Sun varies, with Earth being closer to the Sun in the fall and early winter months. This causes the

Chapter 6: Time, Clocks, and the Stars 53

length of day in apparent solar time to vary by as much as 15 minutes between the Earth's closest position and farthest position from the Sun.

To correct for the length of the day changing over the course of a year, astronomers use mean solar time. Mean solar time is based on apparent solar time averaged over the course of a year. In other words, if you measure the length of each of the apparent solar days in a year, add together the day lengths for all the days in the year, and then divide this by the number of days in a year, the result is the average length of a

solar day, which is mean solar time. A standard watch uses mean solar time that is divided into hours, minutes, and seconds. Global time zones are created using mean solar time.

Mean solar time tells us that it takes 24 hours for Earth to spin once around its axis, which is the length of one mean solar day. But at the same time that Earth is spinning on its axis, it is also rotating around the Sun. Earth travels so fast in its orbit around the Sun that by the time Earth has made one rotation on its axis, it has traveled 2.5 million kilometers in its orbit around the Sun! So, in one mean solar day, Earth has not only rotated once on its axis but has also changed its position in space relative to the Sun.

For example, let's say you are in New York City at noon and you observe that the Sun is at its highest position in the sky. The next day you look at your watch and it is 24 hours later with the Sun once again in its highest position in the sky. On each of these days at noon in mean solar time, the Sun is directly above New York City (the Sun is in its highest position in the sky). The problem is that the Earth, as well as rotating once on its axis, has also moved 2.5 million kilometers in its orbit around the Sun from one noon to the next. Although due to Earth's rotation the Sun is directly above New York City at two noontimes that are 24 hours apart in mean solar time, the *actual* rotation of the Earth (or true rotation) has only taken 23 hours and 56 minutes. Why is this? It's the change of the position of Earth relative to the Sun that makes the length of the mean solar day longer than the length of time for one true rotation. This results in a difference of 4 minutes between how long it takes Earth to make one true rotation on its axis and the length of one day in mean solar time. The length of time it takes for Earth to make one true rotation is called a sidereal day.

You can observe the difference between the length of time of one mean solar day and the length of one sidereal day (one true rotation) if you observe the same stars at the same location in the sky on several consecutive nights. When the stars arrive at the location you've noted, you will discover that the time on your watch will be about four minutes earlier each night. This four minute difference is the result of the difference between mean solar time and the actual rotation time of Earth.

Astronomers developed sidereal time, or star time, to measure time more accurately by using Earth's position relative to distant stars rather than relative to the Sun. The term sidereal comes from the Latin word *sidus* which means "star." Using sidereal time, astronomers are able to calculate the true rotation of Earth. Similar to how

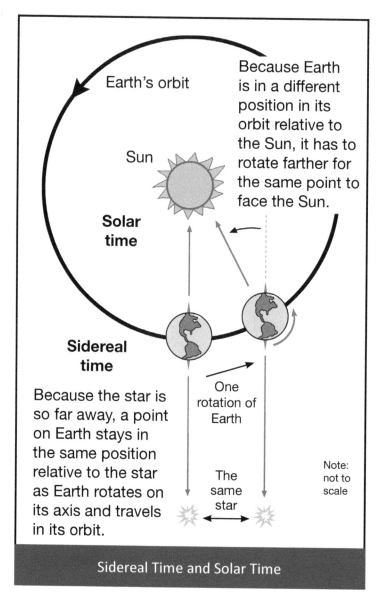

Sidereal Time and Solar Time

a mean solar day is measured by the Sun's highest position in the sky from one day to the next, a sidereal day is measured from the time a distant star appears in its highest position from one night to the next. Because the stars are so far away, Earth's orbit does not affect the position of where the stars appear in the sky — only the rotation of Earth does.

Although sidereal time is a more accurate measure of Earth's rotation on its axis, it is most useful to astronomers. For everyday use, solar time works best.

6.4 Celestial Clocks

In the modern world our days are divided up into segments of hours, minutes, and seconds. Because we no longer rely on observing the location of the Sun to tell time, we use clocks to help us make sure we get to our appointments on time, pick up the laundry on time, eat breakfast, lunch, and dinner on time, and go to bed on time.

Our standard clocks don't take into account the movement of the Moon, other planets, or the stars. A celestial clock, or astronomical clock, on the other hand, is an instrument first built by ancient scholars to provide information about astronomical movements of celestial bodies as well as keeping track of time.

Antikythera Mechanism piece
Courtesy of Marsyas/Wikimedia Commons, CC 3.0 License

Reconstruction of Antikythera Mechanism showing internal gears
Photo by Giovanni Dall'Orto

The first astronomical clock we know of is the Antikythera Mechanism built 2000 years ago by the ancient Greeks. Pieces of the mechanism were discovered by divers in 1900 near the tiny island Antikythera in Greece. Using x-rays to peer into the body of the device and computers to reconstruct how the device worked, scientists have suggested that it was an incredibly accurate astronomical clock able to replicate the irregular motions of the Moon, track the position of the Earth in its orbit around the Sun, and determine the position of Venus, Mars, Jupiter, and Saturn for any chosen date.

Another early astronomical clock was designed by Su Sung of China, built in 1092, and ran until 1126 when the Sung Dynasty was overtaken. This clock is called the Cosmic Engine and is an astronomical clock powered by falling water or falling mercury. The original clock tower was 30 feet tall with a series of interlocking gears rotating with precision. The clock displayed the positions of the Sun, Moon, and planets.

Su Sung's astronomical clock

Prague astronomical clock
Courtesy of Andrew Shiva, CC BY SA 4.0

Astronomical clocks became a spectacle in the European world during the Middle Ages. One of the most famous old astronomical clocks still existing is located in the town square in Prague, the capital of the Czech Republic. The clock was built by Mikulas of Kadan in 1410 and consists of an astronomical dial, a calendar dial, and a window with rotating characters.

Chapter 6: Time, Clocks, and the Stars   57

6.5 Summary

- Mapping the stars is called celestial cartography or uranography.

- Apparent solar time is the time it takes for the Earth to complete one rotation around its axis (complete one day) based on the position of the Sun.

- Mean solar time is the average of apparent solar time over a full year.

- Sidereal time is the actual measurement of Earth's rotation around its axis based on the position of a fixed star.

- Celestial clocks record time plus movements of celestial bodies.

- A star atlas is like a road map of the night sky, mapping the locations and brightness of stars, planets, and other celestial bodies.

6.6 Some Things to Think About

- How do you think star maps have changed over time?

- How do you think star maps help us understand the universe?

- Why do you think ancient people recognized many of the same constellations that we do today?

- What information can you find in a detailed star atlas?

- Why do you think you need different star maps for different times of the year and different times of night?

- Do you think you would use the same star atlas in both the Northern and Southern Hemispheres? Why or why not?

- Why do you think astronomers prefer to use sideral time in their research?

- Do you think an astronomical clock could be helpful to astronomers? Why or why not? Would you like to have one?

Chapter 7 Our Solar System

7.1	Introduction	59
7.2	Planetary Position	59
7.3	Planetary Orbits	60
7.4	Asteroids, Meteorites, and Comets	62
7.5	Habitable Earth	65
7.6	Summary	66
7.7	Some Things to Think About	67

Chapter 7: Our Solar System

7.1 Introduction

In a previous chapter we examined the eight planets of our solar system. We saw that the planets are divided into two broad categories: terrestrial planets and Jovian planets. We discovered that the four planets closest to the Sun are terrestrial planets made mostly of rock, like Earth, and the four outer planets are Jovian planets made mostly of gases, like Jupiter.

In this chapter we will take a closer look at our solar system. A solar system is a group of celestial bodies and the one or more suns they orbit. Our solar system has eight planets orbiting a single sun.

7.2 Planetary Position

If we look at our entire system of planets, we see that the Sun is in the center of the solar system with the planets orbiting the Sun in a particular order. Mercury orbits closest to the Sun followed by Venus, Earth, Mars, Jupiter, Saturn, Uranus, and finally Neptune.

Because the distance from the Sun to the planets is very large, astronomers measure planetary distances in units called astronomical units, or AU. One AU is equal to 149,597,870.7 kilometers (92,955,801 miles). To get an idea of just how far one AU is, imagine that you had to drive from the Earth to the Sun (1 AU) in your car going 97 kilometers per hour (60 miles per hour). To get to the Sun this way, it would take 1,549,263 hours or 64,552 days, or about 177 years!

Using AU to measure the distance of the planets from the Sun, you can see that the four terrestrial planets are relatively close together. All of the terrestrial planets are less than 2 AU from the Sun, with Mercury the closest at 0.387 AU and Mars the farthest at 1.524 AU.

Planets and their distances from the Sun — Courtesy of NASA

In between the terrestrial planets and the Jovian planets is a huge 4 AU space. Jupiter, the closest of the Jovian planets, is 5.2 AU from the Sun, and Neptune, the farthest of the Jovian planets, is an incredibly far 30 AU from the Sun!

7.3 Planetary Orbits

An orbit is defined as the gravitational curved path of one celestial body moving around another celestial body. In other words, the orbit is the "road" a planet travels as it circles the Sun, and the Sun's gravity is what holds the planet in its orbit.

All of the planets orbit the Sun in a counterclockwise direction, and if we take a look straight down at the planetary orbits, we discover that the orbits look almost circular. They are not fully circular and so are technically elliptical, but they are not as elliptical as many people think they are.

One common misconception about Earth's seasons is that it is Earth's orbit that gives us the summer and winter months. However, by examining Earth's orbit it's easy to see that the difference between Earth's farthest and closest distance from the Sun is very small. In other words, as Earth orbits the Sun, Earth's closest position to the Sun is not significantly different from its farthest position from the Sun. The seasons are determined by Earth's tilt on its axis, not its distance from the Sun. One pole of the Earth is tilted toward the Sun in the summer months and away from the Sun in the winter months.

Because there is such a large gap between Mars and Jupiter, astronomers place the planets in two groups. The terrestrial planets make up the inner solar system and the Jovian planets make up the outer solar system.

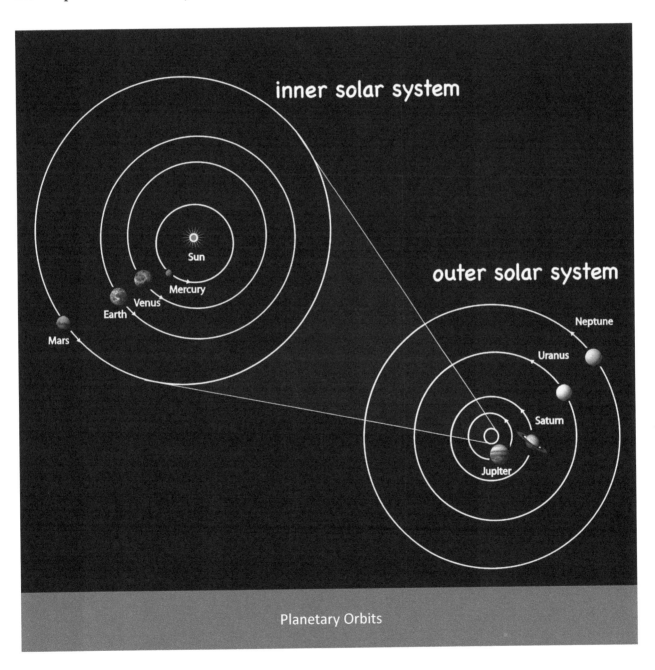

Planetary Orbits

7.4 Asteroids, Meteorites, and Comets

The gap between Mars and Jupiter is not empty space but instead is home to millions of asteroids. The word asteroid comes from the Greek word *aster* which means "star." An asteroid is a small celestial body made mostly of rock and minerals, but when an asteroid is viewed in the sky, it can resemble a small star. However, asteroids are not real stars like our Sun because they are only reflecting light from the Sun rather than emitting their own light. The asteroids between Mars and Jupiter occupy an area known as the Asteroid Belt. Asteroids also exist outside the Asteroid Belt.

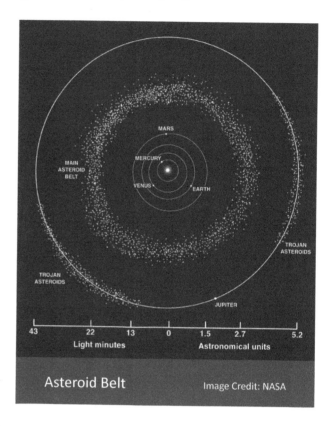
Asteroid Belt
Image Credit: NASA

Asteroid Vesta as seen by NASA's Dawn spacecraft
Courtesy of NASA/JPL-Caltech/UCAL/MPS/DLR/IDA

Scientists estimate that there are 1-2 million asteroids in the Asteroid Belt that are more than 1 km (.62 mi.) in diameter and millions more that are smaller. A few are much larger, like Asteroid Lutetia which is 100 km (62 mi.) in diameter and Asteroid Vesta which is about 525 kilometers (326 mi.) in diameter. Asteroids often have irregular shapes and some have small moons orbiting them.

Chapter 7: Our Solar System 63

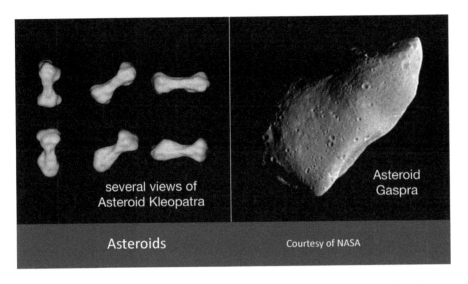

Asteroids — several views of Asteroid Kleopatra; Asteroid Gaspra. Courtesy of NASA

Asteroid Gaspra is an asteroid with an elongated body, and Asteroid Kleopatra has a dog bone shape.

Although there are great distances between asteroids in the Asteroid Belt, asteroids sometimes collide. Because asteroids are moving at great speeds, when they collide, the force of the impact is more than sufficient to shatter rock. Many asteroids have craters on their surface as a result of these high impact collisions.

Asteroids are also found outside the Asteroid Belt and do occasionally impact Earth. Small asteroids, if they cross into the Earth's atmosphere, are called meteors. They often break up into smaller pieces and burn up before reaching the surface of the Earth. Meteors that reach the Earth's surface are called meteorites. Depending on their composition, meteorites are called "stones" or "stony irons."

Meteor over Mauna Kea, Hawaii (the Big Dipper can be faintly seen over the observatory)
Courtesy of Gemini Observatory/Joy Pollard

Scientists are researching asteroids to find out if they contain materials that could be mined in the future. In 2005 the Japan Aerospace Exploration Agency (JAXA) spacecraft Hayabusa landed on the asteroid Itokawa, and in 2010 Hayabusa brought back to Earth a small sample of asteroid dust for analysis. It appears that rather than being solid rock, Itokawa consists of a group of rocks held together by gravity. There are also missions being planned by several countries to see if it is possible to use a controlled impact to change the orbit of an asteroid, the idea being that if an asteroid is headed toward a collision with Earth, it could be deflected so it would miss Earth.

A comet is another type of celestial body found in our solar system. Comets are large chucks of dirty ice. Some comets have an orbit that brings them close to the Sun. When this happens, the Sun's heat vaporizes some of the ice, changing the frozen water and frozen gases directly from the solid state to the gaseous state and creating long tails of gas and dust particles that are visible when the particles reflect light from the Sun.

Two famous comets that can be easily seen when their orbits bring them close to Earth are Halley's Comet and the Hale-Bopp Comet. In 1986 as Halley's Comet passed close to Earth, several spacecraft were able to get close enough to gather information about it. Halley's Comet has a potato-shaped center about 15 kilometers (9 miles) long and a long tail made of various frozen gases such as carbon dioxide, methane, and ammonia. In 1997 Hale-Bopp Comet passed by Earth, displaying a beautiful fluorescent blue-white tail made of ionized carbon monoxide molecules.

Halley's Comet   Hale-Bopp Comet

Comets
Courtesy of NASA

In 2004 the European Space Agency (ESA) launched the Rosetta spacecraft whose mission was to orbit Comet 67P/Churyumov-Gerasimenko and send data back to Earth. It took ten years for Rosetta to arrive at the comet, and Rosetta orbited the comet until the mission ended in 2016. Rosetta also released a lander to the comet's surface, but when it landed, it didn't work. Rosetta collected data as the comet's orbit took it closer to the Sun, enabling scientists to observe the comet as it was "activated" by energy from the Sun, causing the frozen gases to begin to vaporize.

Comet 67P/Churyumov-Gerasimenko as seen by ESA's Rosetta spacecraft

Courtesy of European Space Agency (ESA)/Rosetta NAVCAM
CC BY-SA IGO 3.0

7.5 Habitable Earth

Within our solar system, as far as we know, there are no other planets, moons, or other celestial bodies that can support life as we know it. Scientists have long been searching for other planets like Earth that could be home to extraterrestrial life — life that exists outside the Earth's system. But so far, science fiction novels are the only place extraterrestrial life exists.

What makes Earth uniquely habitable?

One unique feature of Earth is our atmosphere. Our transparent atmosphere helps maintain the necessary balance of water, gas, and energy. No other atmosphere like Earth's has yet been found to exist.

All known life is dependent on liquid water, and the Earth is located at just the right distance from the Sun for liquid water to exist. A little too close and our oceans would boil, leaving no water for life. A little too far away and Earth and our oceans would freeze and be too cold to support life.

The Moon stabilizes Earth's tilt, and the large planets, Jupiter and Saturn, shield the inner solar system from receiving too many impacts by comets. So both the Moon and the planets help stabilize Earth's habitability.

Scientists are using many different space telescopes, probes, and landers to look for planets outside our solar system that are at the right distance from their sun to have liquid water and that might have the other conditions necessary for life as we know it. Within our solar system, scientists think they may have discovered liquid water below the ice on a moon of Jupiter called Europa and a moon of Saturn called Enceladus, but it is not yet known if some form of life exists on either moon. Some scientists think that microbes such as archaea might be able to live in the extreme conditions on these moons.

7.6 Summary

- The terrestrial planets (Mercury, Venus, Earth, and Mars) make up the inner solar system and are "close" to the Sun (less than 2 AU).

- The Jovian planets (Jupiter, Saturn, Uranus, and Neptune) make up the outer solar system, and are "far" from the Sun (more than 5 AU from the Sun).

- Each of the eight planets has a slightly elliptical orbit (very close to circular).

- Asteroids exist throughout the solar system, but most are found in the Asteroid Belt between Mars and Jupiter.

- Earth is the only known habitable celestial body in our solar system and is uniquely suited for life.

7.7 Some Things to Think About

- What do you remember about the planets?

- What is an AU?
  Where does the measurement come from?
  Why do astronomers use it?

- Do you think there's a reason why the terrestrial planets are grouped together and the Jovian planets are grouped together?

  Do you think there's a reason why the small planets are close to the Sun and the big planets are farther away?

  What do you think these reasons might be?

- What is the difference between a comet, an asteroid, a meteor, and a meteorite?

- Why do you think people would want to have a mine on an asteroid?

- What kinds of information do you think scientists learned from the Rosetta mission?

- Do you think if a moon of another planet had liquid water covered by a shell of ice, some kind of life might exist there? Why or why not?

  If so, do you think it would be the same kind of life as on Earth? Why or why not?

- If you were an astronomer looking for life in the universe, where would you look? What would you look for? How would you know if you had found a life form?

Chapter 8 Other Solar Systems

8.1	Introduction	69
8.2	Closest Stars	69
8.3	Brightest and Largest Stars	71
8.4	Planets Near Other Stars	72
8.5	The Circumstellar Habitable Zone	74
8.6	Summary	76
8.7	Some Things to Think About	77

Courtesy of NASA Ames/SETI Institute/JPL-Caltech (artist's concept of Kepler 186f)

8.1 Introduction

In the last chapter we explored our solar system. We saw that the Sun is the center of our solar system, and the planets orbit the Sun in a counterclockwise direction. We discovered that Earth is unique among the planets in our solar system in that it is the only planet that we know to support life. But what about planets in other solar systems? Are there other suns in our universe that have planets orbiting them? Do any of these other solar systems support life? In this chapter we will explore some of our neighboring stars.

8.2 Closest Stars

If we look outside our solar system, we discover that there are countless other stars and solar systems. The closest stars to our solar system are actually part of a triple-star system called the Alpha Centauri system.

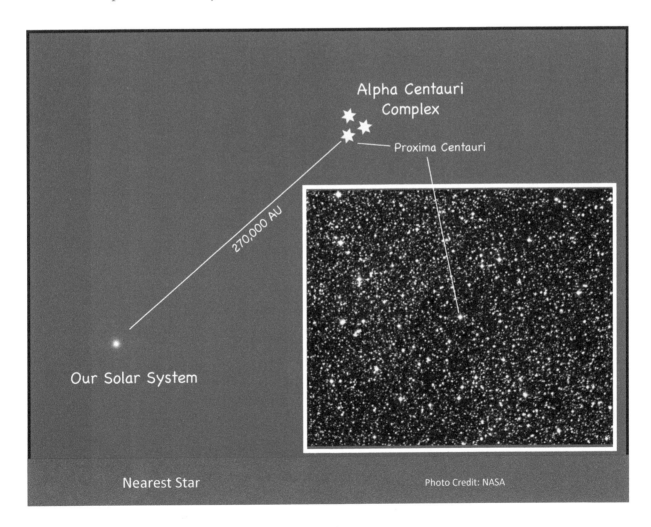

The three stars in this triple-star system are bound together by gravity. Two of these stars, Alpha Centauri A and Alpha Centauri B, are similar to our Sun and orbit each other. The third star, Proxima Centauri, is the star that is closest to Earth, and it orbits Alpha Centauri A and Alpha Centauri B.

Even though Proxima Centauri is closest to our solar system, it is still about 270,000 AU away. This means that the distance from Earth to Proxima Centauri is almost 300,000 times the distance from Earth to our Sun!

The next nearest stars to our solar system are Barnard's Star, Wolf 359, and Lalande 21185. All of these stars are many millions of miles away from our solar system.

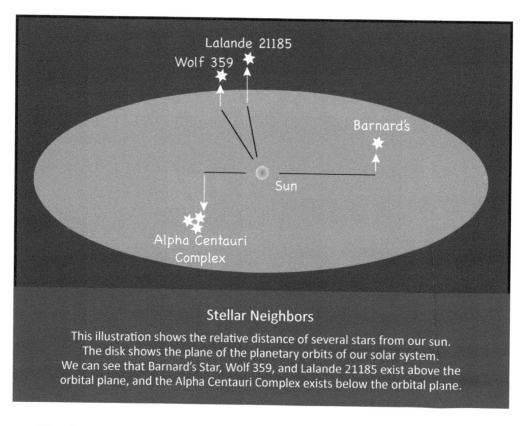

Stellar Neighbors

This illustration shows the relative distance of several stars from our sun. The disk shows the plane of the planetary orbits of our solar system. We can see that Barnard's Star, Wolf 359, and Lalande 21185 exist above the orbital plane, and the Alpha Centauri Complex exists below the orbital plane.

Because stellar distances are extremely large, astronomers measure these distances in parsecs. We saw in the last chapter that Earth is 1 AU from the Sun, which is about 150 million kilometers (93 million miles). A parsec is equal to 206,260 AUs or about 19,000,000,000,000 miles! It is easy to see why astronomers measure stellar distances in parsecs.

Barnard's Star is roughly 1.8 parsecs from Earth, Wolf 359 is about 2.4 parsecs from Earth, and Lalande 21185 is about 2.55 parsecs from Earth.

8.3 Brightest and Largest Stars

The brightest stars in the sky are not necessarily the closest or largest stars. Sirius is the brightest star in the sky, and it is 2.6 parsecs away from our Sun. It is not as close as the Alpha Centauri star system, but Sirius is 20 times brighter than our Sun and over twice as large.

Sirius can be found in the Canis Major constellation. Sirius has a secondary star associated with it called Sirius B which is significantly dimmer than Sirius.

Sirius
Photo Credit: ROSAT, MPE, NASA
Courtesy of Skyview (http://skyview.gsfc.nasa.gov)

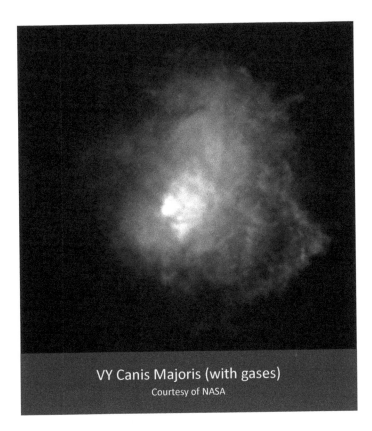

VY Canis Majoris (with gases)
Courtesy of NASA

The largest star visible in the night sky is the star VY Canis Majoris (VY CMa). This star dwarfs our Sun by several magnitudes. A magnitude is the measurement of the brightness of a celestial body.

VY Canis Majoris is considered a red hypergiant star and is located in the constellation Canis Major. VY CMa is very far away from our Sun at a distance of 1500 parsecs. It is a solitary star and does not have multiple stars associated with it. VY CMa is a very active star that emits large amounts of gas during stellar outbursts, which are eruptions of electrically charged particles from a star's surface. These stellar outbursts result in mass being ejected from the star.

8.4 Planets Near Other Stars

Because the large amount of light generated by a star hides the much smaller planets that lie close to the star, it has been difficult to confirm the existence of extrasolar planets. Extrasolar planets (or exoplanets) are planets that orbit stars outside our solar system. Finding exoplanets has been discussed at least since the middle of the 20th century, but it wasn't until 1994 that the existence of planets outside our solar system was confirmed.

Exoplanets as seen by the Hubble Space Telescope (indicated in green)

Courtesy of NASA

Chapter 8: Other Solar Systems 73

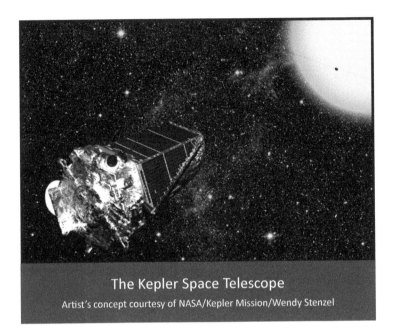

The Kepler Space Telescope
Artist's concept courtesy of NASA/Kepler Mission/Wendy Stenzel

In March 2009 NASA launched the Kepler Space Telescope into an orbit around the Sun with the mission of finding habitable planets around other stars in the Milky Way Galaxy. The Kepler Space Telescope is named after Johannes Kepler (1571-1630 CE), the German astronomer who developed the laws of planetary motion to describe how the planets move around the Sun.

The Kepler Space Telescope uses the transit method to find exoplanets. With this method Kepler detects planets by observing the tiny dimming of a star's brightness as a planet passes in front of, or transits, the star it is orbiting. In order for the transit method to work, the exoplanet's orbit must be aligned with the telescope's line of sight. In other words, Kepler needs to have an edge-on view of the orbit for the telescope to be able see the effect of the planet passing in front of its star. If Kepler is looking at an orbit from the "top," the planet cannot be seen transiting its star. Since a planetary orbit can be oriented at any angle relative to Kepler's point of view, many exoplanets will not be visible to Kepler.

Top: Planet can be detected by Kepler
Bottom: Planet cannot be detected by Kepler

Using data sent to Earth from the Kepler Space Telescope, by November 2016 astronomers had confirmed the existence of over 3400 exoplanets and detected thousands of possible planets. Scientists estimate that in our galaxy alone there may be billions of planets that are of a size similar to that of Earth, and the total number of planets in the Milky Way may be hundreds of billions. It is thought that most of the stars in the Milky Way have planets orbiting them.

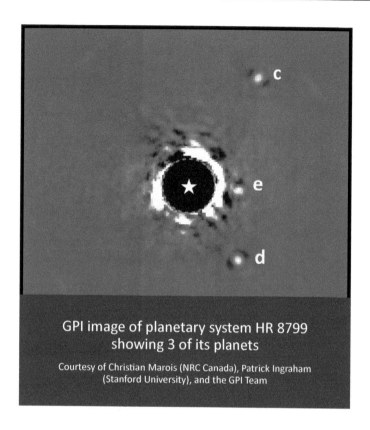

GPI image of planetary system HR 8799 showing 3 of its planets

Courtesy of Christian Marois (NRC Canada), Patrick Ingraham (Stanford University), and the GPI Team

At its Gemini South Telescope in Chile, the Gemini Observatory has installed a new instrument called the Gemini Planet Imager (GPI). The GPI is able to find Jupiter-like planets from an Earth-based observatory rather than from a space telescope. It images planets directly by detecting infrared radiation (heat) from gas planets. The Kepler Space Telescope, on the other hand, uses visible light and the transit method to indirectly detect a planet. Each telescope views only a small portion of the galaxy.

Astronomers classify exoplanets according to their Jupiter-like or Earth-like characteristics. There are massive Jovian-type planets referred to as Jupiters and less massive Jovian planets called Neptunes. There are also planets whose masses are up to 10 times that of Earth, and these are referred to as super-Earths. Exoplanets also differ depending on how far they are from their parent sun. Planets that orbit close to their sun are called hot planets and have extremely high temperatures, while planets farther away are called cold planets due to their colder temperatures.

8.5 The Circumstellar Habitable Zone

Today we know that planetary systems are common in the universe. In order for an exoplanet to support life as we know it, the planet would have to be found in a particular area of its solar system called the Circumstellar Habitable Zone. In this region, an Earth-like planet would be at the right distance from its star to be neither too hot nor too cold to be able to maintain liquid water.

Any given star is surrounded by a Circumstellar Habitable Zone. For small and cooler stars the habitable zone is close to the star, and for larger and hotter stars the habitable zone is farther away.

Chapter 8: Other Solar Systems   75

The Kepler Space Telescope has found terrestrial exoplanets that are about the size of Earth and are in habitable zones, but it is not yet known if any of these planets have conditions that would allow life as we know it to occur.

The search for life on other planets is an exciting area of astronomy to explore. As technology advances, more and more exoplanets will be identified, and we will begin to discover details about their composition, atmosphere, existence of liquid water, and potential for supporting life.

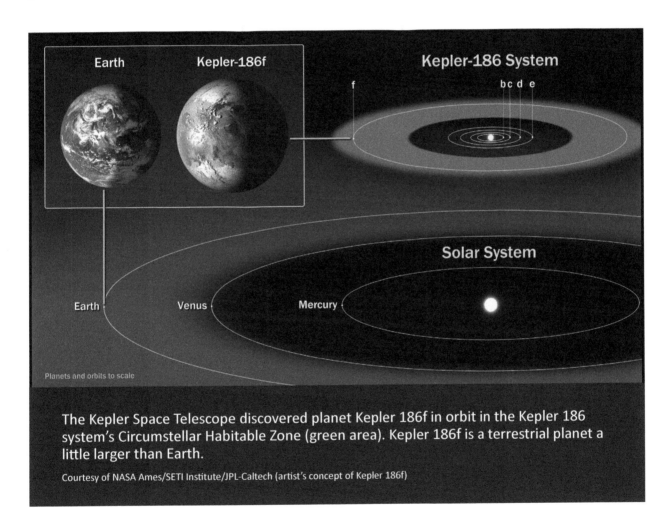

The Kepler Space Telescope discovered planet Kepler 186f in orbit in the Kepler 186 system's Circumstellar Habitable Zone (green area). Kepler 186f is a terrestrial planet a little larger than Earth.

Courtesy of NASA Ames/SETI Institute/JPL-Caltech (artist's concept of Kepler 186f)

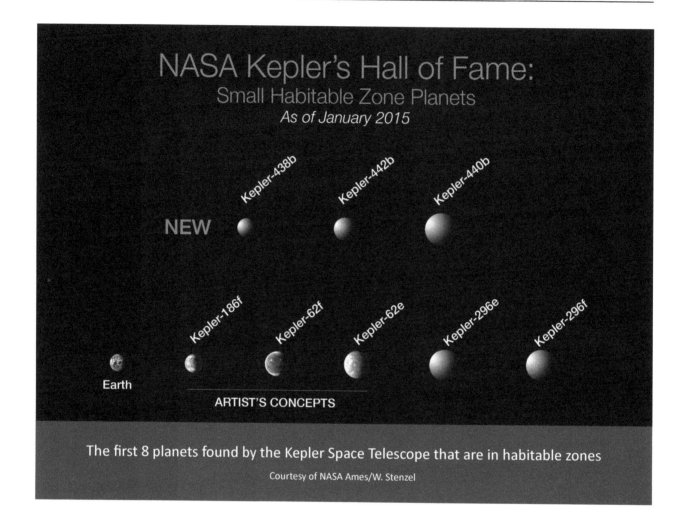

The first 8 planets found by the Kepler Space Telescope that are in habitable zones

Courtesy of NASA Ames/W. Stenzel

8.6 Summary

- Proxima Centauri in the Alpha Centauri system is our nearest stellar neighbor.

- The distances of stars are measured in parsecs. One parsec equals 206,260 AUs.

- The stars that appear brightest and largest from Earth are not the closest stars.

- Extrasolar planets, also called exoplanets, are planets that orbit stars outside our solar system. Many stars are confirmed as having exoplanets.

- In the area called the Circumstellar Habitable Zone, an Earth-like planet would be at the right distance from its star for the presence of liquid water to be possible.

8.7 Some Things to Think About

- Do you think any nearby stars have planets orbiting them? Why or why not?

- If the stars in the Alpha Centauri system have planets, do you think the planets would orbit one star, two stars, or all three? Why?

- What do you think it would be like to live on a planet in a system that had three stars?

- Why do think the brightest star in the sky might not be the largest?

- Do you think the brightness of a star tells an astronomer anything about the star? Why or why not?

- What do you think might happen to a star that is continually emitting large bursts of its gases?

- Do you think that as new technology is developed, scientists will be able to detect more exoplanets? Why or why not?

- Do you think the transit method of exoplanet detection will eventually be replaced by technology that allows for direct imaging of terrestrial exoplanets? Why or why not?

- Do you think any terrestrial exoplanet found in a circumstellar habitable zone will have conditions to support life as we know it? Why or why not?

Chapter 9 Galaxies

9.1	Introduction	79
9.2	Discovering Galaxies	79
9.3	Clusters	83
9.4	At the Galactic Center	84
9.5	Star Formation	86
9.6	Galaxies Interact	86
9.7	Summary	89
9.8	Some Things to Think About	90

Spiral Galaxy M83
Image courtesy of NASA, ESA, and Z. Levay (STScI-AURA)
Acknowledgement: R. Khan (GSFC and ORAU)

9.1 Introduction

We know that Earth is part of a solar system that has eight planets and other objects that orbit a single star, our Sun. Our solar system is traveling through space with billions of stars in a group called the Milky Way Galaxy. Almost all the objects we see in the night sky with the unaided eye are part of the Milky Way Galaxy.

A galaxy is a large group of stars, gas, dust, planets, and other objects in space that are held together by gravitational forces. A galaxy can contain anywhere from a thousand to trillions of stars and can also have remnants of stars that have died, areas of gas and dust where new stars are

Galaxy cluster Abell S0740

Courtesy of NASA, ESA, and The Hubble Heritage Team (STScI/AURA)

formed, and other objects such as black holes. As we observe space through telescopes, we can see faraway galaxies of different shapes and sizes and learn more about their features. Our understanding of galaxies and other objects in space is constantly expanding as new discoveries are made.

9.2 Discovering Galaxies

Prior to the early twentieth century it was thought that everything that could be observed in space was part of our galaxy, the Milky Way. At that time there was no way to measure the distance to very far away celestial objects, although the distance to nearby stars could be measured using parallax. The apparent size and brightness of a star as it is viewed from Earth can't be used to determine how far away it is because how bright a star appears to the eye varies with distance. Therefore, a smaller, dimmer star that is closer to Earth may appear to be brighter than a larger, brighter star that is much farther away. As an example of how distance affects apparent size and brightness, we can compare light from car headlights and light from a small flashlight. The headlights of the car are much brighter and larger than the flashlight. However, if the car is very far away from you and the flashlight is close, it would appear that the flashlight is brighter and larger than the headlights.

The Small and Large Magellanic Clouds
at the VLT (Very Large Telescope), Cerro Paranal, Chile

Courtesy of ESO/B. Tafreshi (twanight.org)

When talking about the light from stars, the total amount of energy a star emits is called the luminosity of the star. The luminosity determines the actual brightness, or absolute magnitude, of the star. The apparent magnitude of a star refers to how bright it appears when viewed from Earth.

In the late 1800s and early 1900s a group of women at the Harvard College Observatory were examining and cataloging the brightness of stars as seen on photographic plates taken using telescopes. Among these women was Henrietta Swan Leavitt (1868-1921 CE) who was studying the photographs of variable stars—stars that grow brighter and dimmer over time. To study these stars, Leavitt compared plates of the same stars that were taken several days to several weeks apart. One type of variable star studied by Leavitt is called a Cepheid star, which is a type of variable star that pulsates, increasing and decreasing in size and brightness.

Leavitt discovered that for Cepheid variable stars expansion and contraction of the star occurs in a regular cycle that can be measured by the number of days it takes the star to go from maximum luminosity through minimum luminosity and back to maximum. This length of time from one maximum luminosity to the next is called the period, which varies from 1 to about 70

A Cepheid star—RS Puppis
surrounded by gas and dust

Courtesy of NASA/ESA/and the Hubble Heritage Team (STScI/AURA)-Hubble/Europe Collaboration; Acknowledgment: H. Bond (STScI & Pennsylvania State University)

days. The Cepheid stars Leavitt was studying were in a fuzzy patch of light called the Small Magellanic Cloud, which at that time was thought to be within the Milky Way Galaxy but was later found to be a small galaxy outside the Milky Way. Leavitt correctly theorized that all the stars within the Small Magellanic Cloud would be a similar distance from Earth, and therefore there would be a relationship between the apparent magnitude and the absolute magnitude of each of these Cepheid stars. Using this theory as a basis, she was able to determine that for Cepheid stars there is a period-luminosity relationship, with a longer period for larger, more luminous stars and a shorter period for smaller, less luminous stars.

Because of the prejudices against women that existed at that time, Leavitt was not allowed to use the resources of the observatory to further research and develop her theories. However, her major discovery allowed other astronomers to calculate the distance to a Cepheid star. By using a measurement of the star's period to calculate its absolute magnitude, astronomers could calculate its distance from Earth. This made it possible for astronomers to make further discoveries that changed how we view the universe.

In the 1920s astronomer Edwin Hubble (1889-1953 CE) was using the telescope at the Mount Wilson Observatory in California, which at the time was the world's most technologically advanced telescope. Hubble was observing fuzzy objects called nebulae that were thought to be clouds of gas within the Milky Way. In particular, he was interested in a patch of light called Andromeda. Hubble discovered a Cepheid star within Andromeda, and using Henrietta Swan Leavitt's theory, he was able to calculate the distance of the star to be about one million light years away—so far away that it had to be outside the Milky Way Galaxy. A light year is the distance light can travel in one year, or 9.46 trillion kilometers, making one million light years a very long distance from Earth. Hubble discovered that Andromeda is a galaxy, and his continued observations of galaxies led him to develop

Some Astronomical Terms

How far? (units of measure in astronomy)

- **astronomical unit** (abbrev., **AU**) • the distance from the Earth to the Sun; equal to 149.6 million kilometers
- **light year** (abbrev., **ly**) • the distance light can travel in one year; equal to 9.46 trillion kilometers
- **parsec** (abbrev., **pc**) • .086 x 10^{13} km (almost 31 trillion km); equal to 3.2 light years
- **kiloparsec** (abbrev., **kpc**) • 1,000 parsecs

How bright?

- **luminosity** • the total amount of energy emitted by a star
- **absolute magnitude** • the actual brightness of a star
- **apparent magnitude** • how bright a star appears to be when viewed from Earth

a system to classify them according to their shape—spiral, barred spiral, elliptical, or irregular. We will learn more about the shapes of galaxies in Chapter 11.

Examples of galaxy shapes (left to right): Spiral galaxy, elliptical galaxy, irregular galaxy

Image credits: Spiral Galaxy NGC 628 (M74)—Gemini Observatory; Elliptical Galaxy M60—NASA, ESA, CXC, and J. Strader (Michigan State University); Irregular Galaxy NGC 1427A—European Southern Observatory (ESO)

Hubble's discovery that outside the Milky Way there are entire galaxies containing huge numbers of stars helped us gain some understanding of the vastness of the universe and to see the universe and our place in it in a different way. We began to see that our solar system is an even smaller part of the universe than previously thought. Advances in technology, such as space telescopes and ground-based radio and optical telescopes, allow us to see farther and farther into space, finding a seemingly endless number of galaxies. Instead of discovering the end of the universe, astronomers keep finding more and more objects farther and

Galaxy EGS-zs8-1 is one of the farthest away galaxies found—over 13 billion light years away.

Hubble Space Telescope image. Arrow shows the location of the galaxy seen enlarged in the inset.

Courtesy of NASA, ESA, P. Oesch and I. Momcheva (Yale University), and the 3D-HST and HUDF09/XDF Teams

farther away. It is unknown exactly how many galaxies exist in the universe, but astronomers estimate there are hundreds of billions or even trillions.

9.3 Clusters

Using powerful telescopes, astronomers are able to observe many features of individual galaxies. They have discovered that rather than being spread out evenly throughout the universe, galaxies are found in groups called galaxy clusters, which may contain anywhere between a few galaxies and hundreds of them. Most galaxy clusters can be further grouped into superclusters that can contain thousands of galaxies.

With the Hubble Space Telescope astronomers have discovered a massive galaxy supercluster called Abell 1689, which is 2.2 billion light years from Earth.

Part of galaxy supercluster Abell 1689

Courtesy of NASA, ESA, the Hubble Heritage Team (STScI/AURA), J. Blakeslee (NRC Herzberg Astrophysics Program, Dominion Astrophysical Observatory), and H. Ford (JHU)

One reason astronomers are very interested in Abell 1689 is because this supercluster contains a huge number of globular clusters, which are dense, globe-shaped clumps of hundreds of thousands of very old stars. Many of these stars are estimated to be at least 10 billion years old. It is thought that little or no star formation is occurring in globular clusters and that all the stars in a globular cluster formed at about the same time.

Studying different globular clusters can help scientists understand how stars

Globular cluster M80 (NGC 6063) contains hundreds of thousands of stars

Courtesy of NASA and The Hubble Heritage Team (AURA/STScI)

and galaxies form and how they change over time. Observing different globular clusters can provide information about how rapidly stars were formed at different stages in a galaxy's development and the chemical composition of the galaxy during formation.

9.4 At the Galactic Center

In most galaxies all the celestial bodies rotate around a central point called the galactic center. Astronomers currently think that almost all galaxes have a black hole at their center and everything in a galaxy orbits this central black hole. Scientists are still learning about black holes. Some theorize that a black hole is an area in space in which a large amount of matter has been compressed into a tiny space, creating extremely strong gravitational forces.

Gases glow at the Galactic Center of Galaxy M106
Courtesy of NASA, ESO, NAOJ/Giovanni Paglioli; Assembling and processing by R. Colombari and R. Gendler

A supermassive black hole emits a jet of matter
Artist's concept courtesy of NASA/JPL-Caltech

The size of a black hole is generally related to the size and mass of the galaxy it is in, with larger galaxies having larger black holes. The black hole at the center of a large galaxy is called a supermassive black hole because it contains more than 1 million times the mass of our Sun.

Scientists think that because black holes are so dense and have so much gravity, light waves entering the black hole cannot escape. This makes black holes effectively invisible — no electromagnetic waves can be detected coming from them. However, scientists can "see" black holes by observing the behavior of stars, gases, and other materials that surround them. By analyzing electromagnetic waves such as X-rays and gamma rays emitted by these other objects and investigating how these objects are affected by a black hole, scientists can draw conclusions about what black holes are and how they work. Studies show that when stars and other objects move too close to a black hole, they are drawn into it and consumed. Some of the particles from these objects are later ejected by the black hole and can be seen as jets of matter coming from it. These jets can be studied to gain a further understanding of black holes.

An artist's representation of a black hole
Matter is orbiting and being pulled into the black hole

Courtesy of Ute Kraus, Physics education group Kraus, Universität Hildesheim, Space Time Travel, (background image of the milky way - Axel Mellinger) CC BY SA 2.5

The supermassive black hole at the center of the Milky Way Galaxy is called Sagittarius A* (pronounced "Sagittarius A star"; also called Sgr A*, pronounced "Saj A star"). Sagittarius A* is thought to have a mass that is 4 million times that of our Sun, and all this mass is tightly packed into an object that only has the diameter of our Sun!

9.5 Star Formation

Stellar nurseries are massive clouds of gas and dust within galaxies where stars are born. Turbulence within these gas and dust clouds causes the materials within them to move and spin, forming dense areas, or knots, of mass. When a knot of gas and dust has gained enough mass, it begins to collapse inward under its own gravity. This creates extreme pressure and causes the material at the center to heat up, forming a protostar, which is the early stage of a star. As the knot of gas and dust continues to collapse, its hot core accumulates more dust and gas. Some of this material will be added to the forming star; some of it may form planets, asteroids, or comets; and some of it may remain as gas and dust. When a star is fully formed, nuclear reactions in the interior create energy that flows outward, keeping the star from collapsing. The formation of a star occurs over a very long period of time. Scientists estimate that a star the size of our Sun would take about 50 million years to become fully formed.

A star forming area in the Eagle Nebula (M16, NGC 6611)

Courtesy of NASA, ESA, and The Hubble Heritage Team (STScI/AURA)

9.6 Galaxies Interact

Before advanced technologies were developed, scientists thought that the universe was calm and quiet with all the stars fixed in place and galaxies staying the same size and shape. Today we know that space is dynamic with rapidly moving objects, stars being born and dying, and galaxies forming and breaking apart. Many objects in space travel together in communities such as galaxies, galaxy clusters, and star clusters, and these communities of celestial objects can interact.

Halton C. Arp (1927-2013 CE) was an American astrophysicist who studied and cataloged unusually shaped galaxies. He wanted to know more about galaxies, how they form and change over time, and why most galaxies have a spiral or elliptical shape. In 1966 Arp published his reference guide *Atlas of Peculiar Galaxies* which contains 338 of the most unusual of the irregularly shaped galaxies grouped by similar characteristics. Galaxies or galaxy groups included in this atlas have the name "Arp" followed by a number. By studying the galaxies in the atlas and other peculiar galaxies, scientists have concluded that their odd shapes result from interactions between galaxies.

Some interacting galaxies seen by the Hubble Space Telescope

Courtesy of NASA, ESA, the Hubble Heritage Team (AURA/STScI)-ESA/Hubble Collaboration, and A. Evans (University of Virginia, Charlottesville/NRAO/Stony Brook University)

NGC 4676, The Mice — a pair of merging galaxies that will eventually form one galaxy

Courtesy of NASA, H. Ford (JHU), G. Illingworth (UCSC/LO), M.Clampin (STScI), G. Hartig (STScI), the ACS Science Team, and ESA

Galaxies are moving through space and begin to interact as they get closer to each other. The tugging and pulling of gravitational and magnetic forces can change the shape of galaxies that are passing each other, sometimes resulting in unusual and beautiful forms. Also, a large galaxy can pull material from a smaller galaxy, tearing apart the smaller galaxy. The celestial

Arp 273 (UGC 1810 top and UGC 1813 bottom)
The galaxy at the top is being distorted into a rose shape by gravitational forces from the galaxy below, which is also being distorted.

Courtesy of NASA, ESA, and the Hubble Heritage Team (STScI/AURA)

Arp 147
It is thought that the galaxy on the left passed through the galaxy on the right, causing a ring shaped structure to be formed.

Courtesy of NASA/STScI

The ring galaxy AM 0644-741
It is thought that the ring was formed when another galaxy passed through this one.

Courtesy of Hubble Heritage Team (AURA / STScI), J. Higdon (Cornell) ESA, NASA

objects pulled from the smaller galaxy can be incorporated into the larger galaxy, increasing its size. Astronomers consider most galaxy interactions to be short-lived, taking about a few hundred million years to complete. This is a short period of time when compared to the life of a typical galaxy, which is estimated to be about 10 billion years.

Sometimes, instead of passing by each other, galaxies collide and merge, or one galaxy may pass through another. It is thought that the merging of galaxies provides energy that increases star production. The large galaxies we see may be the result of many mergers over the life of the galaxy.

Chapter 9: Galaxies 89

Arp 142, The Bird, or The Penguin and Egg

The penguin shaped galaxy was once a flat, spiral disk, but gravitational forces from the elliptical galaxy below it have warped it into a distorted shape.

Courtesy of NASA, ESA, and the Hubble Heritage Team (STScI/AURA)

NGC 4038-4039, The Antennae Galaxies

It is thought that the energy released during the merger of these two galaxies will cause billions of stars to be formed.

Courtesy of NASA, ESA, and the Hubble Heritage Team (STScI/AURA)-ESA/Hubble Collaboration, Acknowledgment: B. Whitmore (Space Telescope Science Institute [STScI])

9.7 Summary

- A galaxy is a large group of stars, gas, dust, planets, and other objects in space that are held together by gravitational forces.

- For Cepheid variable stars there is a period-luminosity relationship which makes it possible to measure their distance from Earth.

- Galaxies are found in groups called galaxy clusters, which may contain anywhere from a few galaxies to hundreds of them.

- Globular clusters are dense, globe-shaped clumps of hundreds of thousands of very old stars thought to be among the oldest stars in the universe.

- Astronomers currently think every large galaxy has a black hole at its center and everything in the galaxy orbits the black hole.

- Stars are formed from matter found in massive clouds of gas and dust in galaxies.

- The gravitational fields of galaxies begin to interact as the galaxies move closer to each other, changing their shape and composition.

9.8 Some Things to Think About

- In the photograph in Section 9.1, how many different shapes of galaxies can you see? How would you describe them?

  Do you think all the spots of light in the photograph are galaxies? Why?

- How do you think our ideas about the universe change as new and more powerful telescopes are developed?

- Why do you think galaxies and stars form clusters?

A star forming area called R136 in the 30 Doradus Nebula. The blue blobs are hot, young stars.

Courtesy of NASA, ESA, and F. Paresce (INAF-IASF, Bologna, Italy), R. O'Connell (University of Virginia, Charlottesville), the Wide Field Camera 3 Science Oversight Committee, and the Hubble Heritage Team (STScI/AURA)

ESO 593-IG 008, or IRAS 19115-2124, also called The Bird or Tinker Bell Galaxy

Three galaxies are merging to form this peculiar galaxy.

Images courtesy of European Southern Observatory (ESO); taken with the ground based Very Large Telescope, Cerro Paranal, Chile; visible and infrared parts of the electromagnetic spectrum are shown

- Why do astronomers like to study globular clusters?

- Where do you think the name black hole comes from?

- How do astronomers gather data about black holes?

- How would you describe how new stars are formed?

- What forces cause interacting galaxies to change their shape?

Chapter 10 Our Galaxy— The Milky Way

10.1	Introduction	92
10.2	Shape and Structure	93
10.3	Size	97
10.4	Viewing the Milky Way	99
10.5	Summary	103
10.6	Some Things to Think About	103

Illustration courtesy of NASA/JPL-Caltech/R. Hurt (SSC Caltech)

10.1 Introduction

The Milky Way seen above the telescopes at ESO's Paranal Observatory in Chile
Courtesy of G. Hüdepohl (atacamaphoto.com)/ESO (European Southern Observatory)

Recall that a galaxy is a large collection of stars, gas, dust, planets, and other objects held together by its own gravity, and the Milky Way Galaxy is the galaxy that our Earth, Sun, and entire solar system reside in.

We cannot observe the entire Milky Way Galaxy because Earth is one very tiny speck in the midst of its vast area. We live inside the galaxy, and the distance is too great for us to be able to travel outside the Milky Way to view it as a whole. This makes it very challenging to study. We might compare the effort to map our own galaxy to a scuba diver trying to map the entire Earth from

Milky Way galactic center as seen in infrared wavelengths (see Section 19.4) — showing stars, star clusters, glowing hydrogen gas, sculpted areas formed by winds from hot, massive stars

Courtesy of NASA, ESA, Q. D. Wang (University of Massachusetts, Amherst) and NASA, Jet Propulsion Laboratory, and S. Stolovy (Spitzer Science Center/Caltech)

within one small part of the ocean. Studying other galaxies that are outside the Milky Way helps us understand more about our own galaxy. It is easier to gather significant details about these other galaxies because we can observe the whole structure of the galaxy.

10.2 Shape and Structure

Early astronomers had very different ideas about the size, shape, and nature of the Milky Way than we do today. In 1785 the English astronomer William Herschel (1738-1822 CE) presented a paper with his conclusions about the size and shape of our galaxy. With the aid of a telescope and assisted by his sister Caroline Herschel (1750-1848 CE), William counted and mapped as many stars as could be seen in every direction. By estimating the distance to the several hundred stars that could be observed at that time and recording their location, a map was drawn that was thought to represent our galaxy's shape and size. Herschel's map showed the Milky Way as a flat, disk-shaped collection of stars with our Sun at or near the center. We now know that the Milky Way is not entirely flat nor is the Sun at its center, but Herschel was correct in predicting that it has a flattened disk shape.

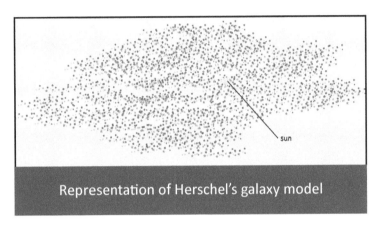
Representation of Herschel's galaxy model

Astronomers estimate that there are billions or trillions of galaxies outside the Milky Way, and with the help of ground-based and space-based telescopes they have found galaxies of many shapes and sizes. Recall that astronomers group the many variations they see into several basic types of galaxies: spirals, barred spirals, ellipticals, and irregulars. We will learn more about other galaxies in the next chapter.

NGC 6744 - Astronomers think if we could see the Milky Way from above, it would look like this galaxy.
Courtesy of European Southern Observatory (ESO), La Silla Observatory, Chile

By using various mapping methods and different instruments, astronomers have concluded that the Milky Way Galaxy contains hundreds of billions of stars, enough gas and dust to form billions more stars, and probably billions of planets. The Milky Way is most likely a barred spiral galaxy. If we could leave our galaxy in a spaceship and look at the Milky Way from above, we would expect to see spiral arms similar to a pinwheel, with these spiral arms extending from the ends of an elongated bar-shaped structure made of bright stars at the galactic center. Most new star formation occurs within the spiral arms.

Astronomers have concluded that the Milky Way has two major spiral arms, Scutum-Centaurus and Perseus, that are packed with stars, gas, and dust. The two minor arms, the Norma Arm and the Sagittarius Arm, have as much gas and dust but fewer stars. Our solar system resides on a small, partial spiral arm called the Orion Spur located between the Sagittarius Arm and the Perseus Arm.

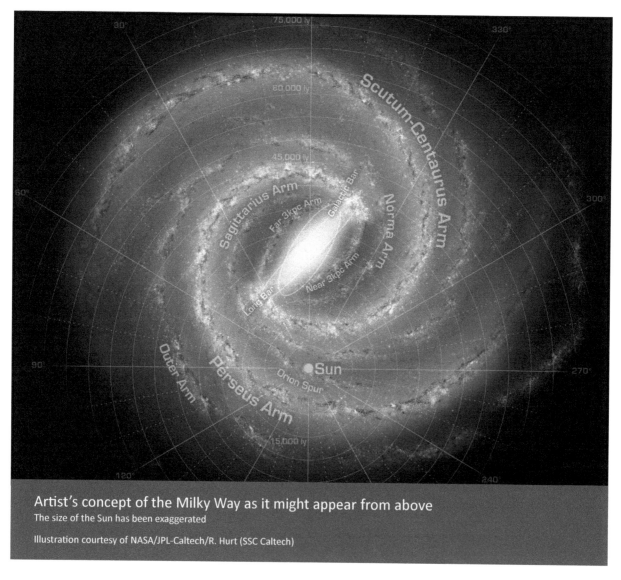

Artist's concept of the Milky Way as it might appear from above
The size of the Sun has been exaggerated

Illustration courtesy of NASA/JPL-Caltech/R. Hurt (SSC Caltech)

The Milky Way Galaxy can be divided into four main regions. The galactic bulge (also called the central bulge), the thin disk, the thick disk, and the galactic halo.

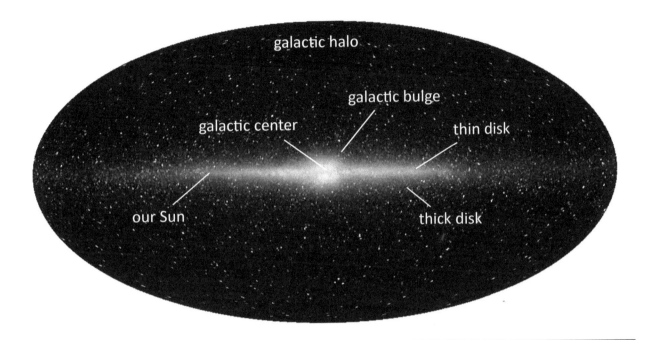

Milky Way Galaxy — edge-on view
Infrared wavelength image taken by COBE satellite
Courtesy of E. L. Wright (UCLA), The COBE Project, DIRBE, NASA

The Milky Way's galactic bulge is a thick, elongated bar-shaped area of stars, dust, and gas orbiting the galactic center. Because there is so much dust and gas in the galactic bulge, visible light from inside the galactic bulge is blocked from our view. However, instruments that detect infrared, radio waves, and X-rays can be used to explore the bulge. It is thought that from an edge-on view the galactic bulge would appear to have a shape similar to a peanut in the shell. It is extremely dense with old stars, and star formation may be taking place in the inner region of the bulge.

The peanut shaped galactic bulge
Artist's concept courtesy of ESO/NASA/JPL-Caltech/M. Kornmesser/R. Hurt

The Arches Cluster
The blue stars are hot and young, the red stars are cooler and old.
Courtesy of NASA/ESA/Hubble Space Telescope

Several very dense globular clusters of stars have been found in the galactic bulge. The Arches Cluster is the densest of these and also the closest to the galactic center. In an area within the Arches Cluster equal to the distance from the Sun to our nearest star, we would see over 100,000 stars. The Arches Cluster contains over 150 stars that are the brightest and most massive in the Milky Way Galaxy.

The supermassive black hole Sagittarius A* lies at the center of the galactic bulge. At 4 million times the mass of the Sun, Sgr A* is a rather small black hole and generally not very active. However, it does have some outbursts of flares that can be picked up with instruments that detect X-rays. It is thought that these flares are extra energy burped out by the black hole as it consumes fuel such as gas or possibly asteroids. A cloud of very hot gas with temperatures of up to about 1000° C has been detected spiralling around Sgr A* and may be feeding it.

Galactic center imaged in infrared, Sgr A* flares imaged in X-ray
Courtesy of NASA, JPL-Caltech (NuSTAR Project)

When we see the Milky Way as a faint band of light in the night sky, we are viewing the thin disk edge-on from Earth's position within it. The thin disk orbits the galactic bulge, contains the spiral arms and the majority of the stars in the galaxy, and would look almost circular when viewed from above. Our Sun and the Earth are located in the thin disk. Stars of all ages are found here and clouds of gas and dust create beautiful nebulae and star forming areas. Most of the youngest stars are found here. The thick disk surrounds the thin disk and contains mostly older stars. The thin and thick disks together are called the galactic disk or the stellar disk.

Milky Way star density map showing part of the thin and thick disks; brightest areas have the most stars, dark areas are thick clouds of dust and gas; Large and Small Magellanic Clouds are below the Milky Way

Courtesy of ESA/Gaia, License CC BY-SA 3.0 IGO; Acknowledgement: Image preparation by Edmund Serpell at ESA's European Space Operations Centre in Darmstadt, Germany

Surrounding the entire galaxy is the galactic halo, a spherical grouping of gas, individual old stars, and globular clusters. Over 150 globular clusters of stars that are over 10 billion years old have been discovered in the Milky Way with most being found in the halo and fewer in the thick disk. The gaseous part of the halo is enormous and is estimated to be at least 600,000 light years in diameter. It is made up of warm and very hot gases with temperatures estimated at between 100,000 and 2.5 million degrees C. There is little or no dust in the halo. The Large Magellanic Cloud and Small Magellanic Cloud galaxies are within the Milky Way's halo.

Gas halo around the Milky Way Galaxy

Illustration courtesy of NASA/CXC/M. Weiss; NASA/CXC/Ohio State/A. Gupta et al.

10.3 Size

How big is the Milky Way Galaxy? Since we are on a small planet circling a relatively small sun off to one side of a huge galaxy, it's tough to measure its size. Because we can't take a spaceship and a tape measure to the end of the galaxy, astronomers estimate distances by using other techniques. Early astronomers assumed that all stars are equally bright and used this assumption to estimate distances to faraway stars. With more advanced technology, astronomers are able to get a more accurate picture of objects and distances in the galaxy.

In the early 20th century an American astronomer named Harlow Shapley (1885-1972 CE) studied globular clusters within the Milky Way. Shapley was able to calculate the distances to the different globular clusters by using the period-luminosity relationship of Cepheid stars and RR Lyrae stars. Like Cepheid stars, RR Lyrae stars are variable stars, but they have a short period of somewhere between four hours and one day, and they are more numerous in the galaxy than Cepheids. Shapley found that globular clusters in the halo are arranged in a spherical grouping, and he concluded that the center of the galaxy would be at the center of the spherical group. By finding the center of the galaxy and using RR Lyrae stars to measure distances, Shapley was able to observe that the Sun is not at the center of the Milky Way. His estimate of the Sun being 50,000 light years away from the galactic center was not accurate, but his discovery that the Sun is far from the galactic center was a major change in how the universe was viewed at that time.

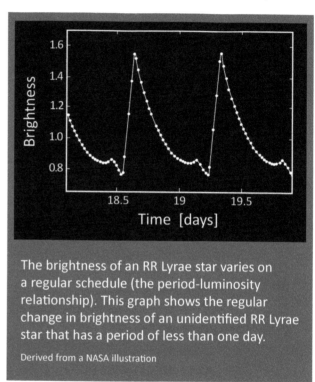

The brightness of an RR Lyrae star varies on a regular schedule (the period-luminosity relationship). This graph shows the regular change in brightness of an unidentified RR Lyrae star that has a period of less than one day.

Derived from a NASA illustration

Astronomical Math!

Measuring Distance to Stars

Basic Formula

$$\text{distance}^2 = \text{absolute magnitude (luminosity)}/\text{apparent magnitude}$$

The *absolute magnitude* of Cepheid and RR Lyrae stars can be calculated using the period-luminosity relationship, and the *apparent magnitude* is measured from Earth. These values can then be plugged into the equation to find distance. From this formula it can be seen that the farther away a star is, the dimmer it will appear when viewed from Earth.

Astronomers use different units of measure to express distances in space. Recall that a light year (ly) is the distance that light travels in one year, and one light year is 9.4607 x 10^{12} km (9.46 trillion kilometers). Another unit of measure for astronomical distances is the parsec (pc). A parsec is 3.086 x 10^{13} km (almost 31 trillion km), or 3.2 light years—a very big number! Astronomers also use a unit of measure called the kiloparsec (kpc) which is equal to 1,000 parsecs. For some time astronomers have estimated the diameter of our galaxy to be about 100,000 light years in diameter. This diameter can be expressed as 100,000 light years, 30,000 parsecs, or 30 kiloparsecs—these all mean the same distance. Whichever measurement system is used, you can see that the Milky Way Galaxy is unimaginably huge!

10.4 Viewing the Milky Way Galaxy

When we talk about observing objects in space, we often think about using our eyes aided by telescopes. Recall that what we see with our eyes is the part of the electromagnetic spectrum that is referred to as visible light, or the visible spectrum.

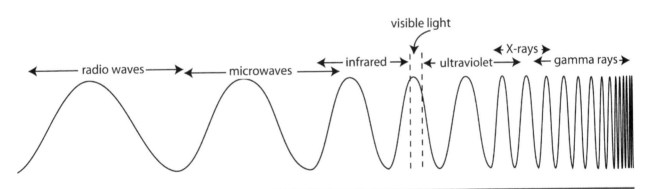

Electromagnetic spectrum

Visible light is only a very small part of the electromagnetic spectrum. By using different types of instruments, scientists can collect data from parts of the electromagnetic spectrum that we cannot see with our eyes. For example, if you look at your hand, you can see the skin and fingernails that are on the surface. If you have an X-ray taken, you will be able to see the bones underlying the skin. These are two different ways of seeing the same object (your hand), one using visible light that is detected by the eyes and the other using X-rays that are detected mechanically. (See *Focus On Middle School Physics* for more about the electromagnetic spectrum).

Looking at the diagram of the electromagnetic spectrum, you can see that the peaks of the longer radio waves are farther apart, or occur less frequently, than the peaks of the shorter gamma ray waves which are closer together and occur with greater frequency. Since the length of the wave (distance between two peaks) determines the frequency at which the wave occurs, wavelength and frequency are often used interchangeably. If you hear that gamma rays are short wavelength or that gamma rays are high frequency, these mean the same thing.

Celestial bodies emit, or radiate, waves of different wavelengths. These emitted waves are referred to as electromagnetic radiation. Different types of instruments are used to detect different wavelengths of the electromagnetic spectrum, and a group of wavelengths is called a spectral band. For example, our eyes detect wavelengths that are in the visible spectral band (or visible spectrum).

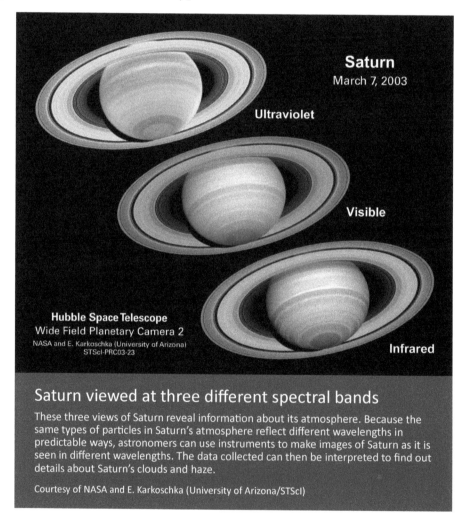

Saturn viewed at three different spectral bands

These three views of Saturn reveal information about its atmosphere. Because the same types of particles in Saturn's atmosphere reflect different wavelengths in predictable ways, astronomers can use instruments to make images of Saturn as it is seen in different wavelengths. The data collected can then be interpreted to find out details about Saturn's clouds and haze.

Courtesy of NASA and E. Karkoschka (University of Arizona/STScI)

The wavelength data collected is transformed by computers into images that scientists can study. Images made from wavelengths outside the visible light spectral band are called false-color images because colors are assigned by artists and computers according to the data being processed.

Chapter 10: Our Galaxy—The Milky Way

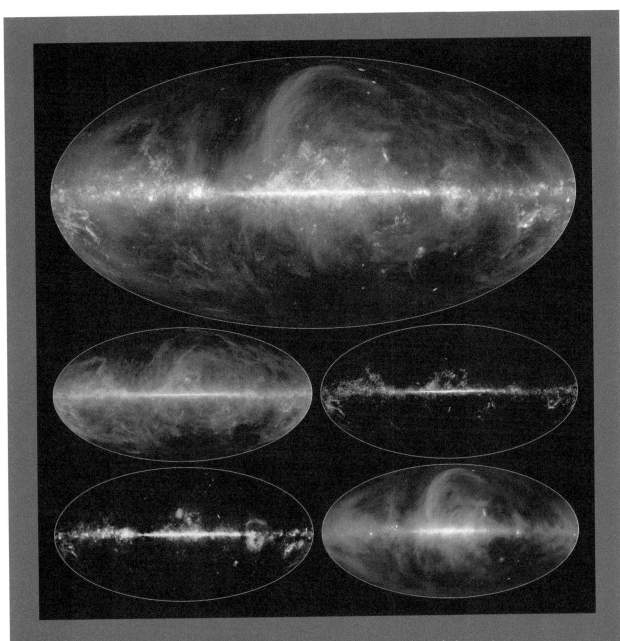

The Milky Way Galaxy seen in different wavelengths of the microwave spectral band of the electromagnetic spectrum.

The large image at the top combines the other four images to make a composite view of the Milky Way. The smaller images below it are the individual parts of the composite. All images are false-color.

Upper left: Dust glow. The red colors show the heat of dust in the galaxy although it is only -253°C.

Upper right: Carbon monoxide gas appears as yellow in the densest clouds of gas and dust where stars are formed.

Lower left: The electrons and protons of hot, ionized gas near massive stars are colored green.

Lower right: The blue shows fast-moving electrons captured in the Milky Way's magnetic field.

Courtesy of ESA/NASA/JPL-Caltech

Great Observatories' Unique Views of the Milky Way
Spitzer Space Telescope • Hubble Space Telescope • Chandra X-Ray Observatory
NASA / JPL-Caltech / ESA / SSC / CXC / STScI
ssc2009-20a

Milky Way Galactic Center

The false-color image at the top is a composite of the images below. By detecting infrared and X-ray wavelengths, we are able to see through the dust of the galaxy to observe the galactic center. The bright spot to the right of center is the galactic core where supermassive black hole Sgr A* resides.

Red - infrared wavelengths detected by the Spitzer Space Telescope. Here we can see hundreds of thousands of stars that heat the surrounding gas and dust making it glow when viewed in infrared. The gas and dust have been formed into blobs, clouds, and stringy structures by radiation and winds from stars.

Yellow - near-infrared wavelengths observed by the Hubble Space Telescope reveal energetic regions of star formation areas of warm gas and hundreds of thousands of stars. Radiation and winds from stars have formed the arcs and other structures seen here.

Blue and violet - X-rays seen by the Chandra X-Ray Observatory. Lower energy X-rays show as pink areas with higher energy X-rays appearing as blue. This view shows gas that has been heated to millions of degrees by stellar explosions, emissions from SgrA*, and winds from massive stars.

Courtesy of NASA, ESA, SSC, CXC, and STScI

10.5 Summary

- The galaxy Earth resides in is the Milky Way Galaxy.

- The Milky Way Galaxy is a barred spiral galaxy with a galactic bulge at the center of a stellar disk.

- Earth resides between the Sagittarius Arm and the Perseus Arm on a small, partial spiral arm called the Orion Arm.

- The size of the Milky Way Galaxy is estimated to be at least 100,000 light years (ly), 30,000 parsecs (pc), or 30 kiloparsecs (kpc) in diameter. These are different ways of stating the same distance in space.

- Astronomers use different instruments to detect electromagnetic radiation that lies outside the visible part of the spectrum, thus allowing them to visualize features of the Milky Way Galaxy that cannot be seen with the eyes.

10.6 Some Things to Think About

- Do you think we will ever be able to see the Milky Way Galaxy from outside it? Why or why not?

- What are the main features of the Milky Way Galaxy? Do you think most galaxies have these features? Why or why not?

- Is our solar system positioned in the Milky Way Galaxy where you would have expected it to be? Why or why not?

- How did Harlow Shapley discover the middle of the Milky Way Galaxy?

- What are three units of measure used by astronomers when talking about distances in space?

- Do you think new theories about the size and structure of the Milky Way Galaxy will be formed as new technologies are developed? Why or why not?

- What objects in the Milky Way Galaxy would we not know about if we had instruments that only detected visible light wavelengths?

Chapter 11 Other Galaxies

11.1	Introduction	105
11.2	Spiral Galaxies	106
11.3	Barred Spiral Galaxies	107
11.4	Elliptical Galaxies	109
11.5	Irregular or Peculiar?	110
11.6	Radio Galaxies	112
11.7	Summary	114
11.8	Some Things to Think About	114

11.1 Introduction

In Chapter 10 we looked closely at our home galaxy, the Milky Way. We discovered that the Milky Way is most likely a barred spiral galaxy with a bulge in the center of an almost circular thin disk of stars. Although until the 1920s it was believed that the Milky Way was the only galaxy in the universe, astronomers now estimate that there are billions of galaxies.

After Edwin Hubble made the exciting discovery that Andromeda is a galaxy outside the Milky Way, he began to observe many other galaxies and found that they come in different shapes and sizes. This led him to develop a classification system that grouped galaxies by how they look.

Galaxy cluster Abell 2744

Courtesy of NASA, ESA, and J. Lotz, M. Mountain, A. Koekemoer, and the HFF Team (STScI)

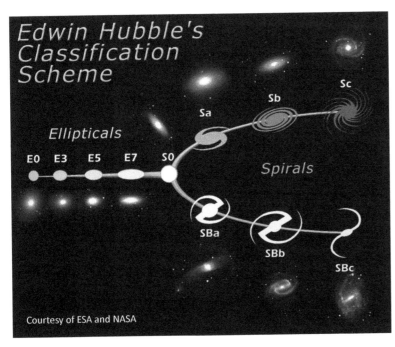

He first grouped the elliptical galaxies that are spherical or oval and don't have other significant features. Next came the spiral galaxies with central bulges and spiral arms. Spiral galaxies were further divided into spirals with a spherical central bulge and spirals with a barred central bulge. Because of the shape of his classification diagram, it's referred to as the Hubble Tuning Fork. Hubble included four major types of galaxies in his system: spiral galaxies, barred spiral galaxies, elliptical galaxies, and irregular galaxies. The irregular galaxies are oddballs with characteristics that vary from galaxy to galaxy so they don't appear on the Hubble Tuning Fork. Although there are other classification schemes, the Hubble classification of galaxies developed in 1926 is still used today as a basic way to sort the variety of galaxies observed by astronomers.

Different theories have been developed to explain how galaxies get their shape. Although the process of galaxy formation is still under debate, it does appear that galactic shapes are a result of natural developmental stages. Also, galaxy mergers and gravitational interactions of galaxies that are passing each other can play a part in shaping a galaxy.

11.2 Spiral Galaxies

Like the Milky Way Galaxy, all spiral galaxies have the same basic features—a pinwheel shape with a central bulge in the middle, spiral arms radiating from the center with the spiral arms forming a flattened galactic disk, and an extended galactic halo around the whole galaxy. Thick clouds of gas and dust in the spiral arms provide the materials necessary for birthing stars, and hot, young stars are found in the spiral arms. These young

The Andromeda Galaxy, the closest galaxy to the Milky Way, is classified as an Sb galaxy
Courtesy of NASA/JPL-Caltech

stars give spiral arms their blue appearance. The greatest number of old, reddish stars are found in the galactic bulge and the halo, and some of these old stars are grouped in globular clusters which are most commonly found in the halo.

In Hubble's classification scheme, spiral galaxies are given the letter S. The letters a, b, and c designate further subdivisions according to the size of the galactic bulge and how tightly wound the spiral arms are. Sa galaxies have the largest galactic bulge and the most tightly wound spiral arms, and Sc galaxies have small galactic bulges and more loosely wound spiral arms.

Spiral Galaxies

Upper left:
Sa galaxy Messier 81 (M81)
Courtesy of NASA/JPL-Caltech/S. Wilmer (Harvard-Smithsonian Center for Astrophysics)

Upper right:
Sb galaxy, The Whirlpool Galaxy M51
M51 is interacting with dwarf galaxy NGC 5195

Courtesy of NASA, ESA, S. Beckwith (STScI), and The Hubble Heritage Team (STScI/AURA)

Left:
Sc galaxy NGC 3938

Courtesy of Adam Block/Mount Lemmon SkyCenter/University of Arizona

Spiral galaxies are easy to spot when they are tilted with respect to Earth because we can see their structure as we view the galaxy from above or at an angle. When observing a galaxy edge-on, astronomers look for features such as a galactic bulge and areas of dust and star formation to determine that the galaxy is a spiral.

11.3 Barred Spiral Galaxies

The second category of galaxies described by Hubble is the barred spiral galaxy. A barred spiral galaxy has the same features as a spiral galaxy except the spiral arms originate at the ends of a bar-shaped region that goes through the center of the galaxy's galactic bulge. Spiral and barred spiral galaxies make up most of the large, bright galaxies that are nearest to Earth. It's estimated that about two-thirds of spiral galaxies are barred. Barred spiral galaxies are designated by the letters SB with the letters a, b, and c designating the size of the galactic bulge and how tightly the spiral arms are wound. The Milky Way Galaxy is grouped with the barred spiral galaxies. Although the size of the bar and the shape of the spiral arms are still uncertain, the Milky Way is generally classified as an SBb galaxy.

Barred Spiral Galaxies

Left:
SBa galaxy NGC 1300
Courtesy of NASA, ESA, and The Hubble Heritage Team (STScI/AURA)

Lower left:
SBb galaxy NGC 1365
Courtesy of NASA, ESA, and the Hubble SM4 ERO Team

Lower right:
SBc galaxy NGC 986
Courtesy of ESA/Hubble & NASA

Sometimes astronomers can't distinguish between spiral galaxies and barred spiral galaxies. This is especially true when the galaxy is edge-on when viewed from Earth and we can only see the galaxy from the side and not the top.

11.4 Elliptical Galaxies

Elliptical galaxies are named for their shape, which varies from spherical to elongated. These galaxies are very different from both spiral galaxies and barred spiral galaxies because ellipticals have no spiral arms or internal features such as a galactic bulge; however, some of them have a supermassive black hole at the galactic center. Elliptical galaxies are thought to be the most numerous type of galaxy and have sizes that cover a very broad range from dwarf ellipticals that are less than one-tenth the size of the Milky Way Galaxy to ellipticals that are the most massive of all galaxies. A giant elliptical can be as much as two million light years in diameter, can contain a trillion or more stars, and may have a supermassive black hole that is three billion times the mass of the Sun!

Elliptical Galaxy IC 2006

Courtesy of ESA/Hubble and NASA
Acknowledgement: JudySchmidt and J. Blakeslee
(Dominion Astrophysical Observatory)
Science Acknowledgement: M. Carollo (ETH, Switzerland)

Elliptical Galaxy NGC 1132

Courtesy of NASA, ESA, and the Hubble Heritage (STScI/AURA)-ESA/Hubble Collaboration
Acknowledgment: M. West (ESO, Chile)

Most elliptical galaxies have a smooth appearance, with a dense concentration of stars at the center gradually decreasing outward. This gives ellipticals the appearance of having a very bright center with the brightness fading outward. Elliptical galaxies contain a lot of older, cooler stars, giving them a reddish color, and it has been theorized that when these stars were created, the gas needed for new star formation was used up, making the galaxies

Visible light and near-ultraviolet light images are combined in this view of elliptical galaxy NGC 4150. The inset shows a close-up of the active core. By imaging elliptical galaxy NGC 4150 in near-ultraviolet light, the presence of new, young stars are revealed as blue areas. Streams of dust are clearly visible.

Courtesy of Courtesy of NASA, ESA, R. M. Crockett (University of Oxford, UK), S. Kaviraj (Imperial College London and the University of Oxford, UK), J. Silk (University of Oxford, UK), Max Mutchler (STScI), Robert O'Connell (University of Virginia), and the WFC3 Scientific Oversight Committee

incapable of birthing new stars. But recent research using ultraviolet light imaging has shown that many ellipticals do contain gas, dust, and young stars. The energy and material needed to make stars in elliptical galaxies is believed to come from the merger of an elliptical with a smaller galaxy.

In the Hubble classification scheme elliptical galaxies are denoted by the letter E and subdivided further with a number from 0 to 7. An elliptical galaxy classified as E0 (pronounced "E-zero") is nearly circular, and an elliptical galaxy designated E7 is the most elongated. Determining the true dimensions of an elliptical galaxy can be a problem because if the end of an egg-shaped galaxy is pointed toward Earth, it will appear to be spherical, and of course we can't travel around it to get a different view.

11.5 Irregular or Peculiar?

The final class of galaxies identified by Hubble are those galaxies referred to as irregular galaxies (designated as Irr). This is a broad term that describes all of those galaxies that don't seem to fit into the categories of spiral or elliptical. Irregular galaxies aren't really spiral, although they may have an arm or two. They aren't really elliptical but can be circular or elongated. Irregulars don't really have a galactic bulge in the center, but they can have a high density of stars off to one side. Sometimes they just look a bit messy. Estimates put irregulars at somewhere around 25% of all galaxies in the universe.

To help better understand irregular galaxies, astronomers have tried to further classify them. This is difficult because they have few common characteristics. However, most irregulars don't have a galactic center, are asymmetrical in shape, and are blue in color because they have active star formation. Irregular galaxies are often categorized as normal irregular galaxies (Irr I) and peculiar irregular galaxies (Irr II).

Normal irregular (Irr I) galaxy NGC 1472A

The blue areas in this image indicate a massive amount of star formation. The energy for the star formation may be arising from NCG 1472A colliding with intergalactic gases or from interaction with forces coming from other nearby galaxies.

Courtesy of NASA, ESA, and The Hubble Heritage Team (STScI/AURA)

Peculiar irregular galaxy (Irr II) M82—the Cigar Galaxy

Some of M82's peculiarity comes from the effects of its passing by the spiral galaxy M81 (not in view). But it is thought that most of the expanding hydrogen gas (shown as red here) is being pushed away from the galaxy by strong stellar winds coming from many stars. These stellar winds combine to form a galactic "superwind." The filaments of gas in M82 spread out to a distance of over 10,000 light years.

Courtesy of NASA, ESA, The Hubble Heritage Team, (STScI/AURA)
Acknowledgement: M. Mountain (STScI), P. Puxley (NSF), J. Gallagher (U. Wisconsin)

Irr I galaxies are the normal irregulars which tend to be small at 1-10 kpc in diameter. Some may be seen to have a hint of internal structure, such as a piece of a spiral arm.

Irr II galaxies are peculiar irregulars that look distorted or misshapen in some respect. There isn't really a clear distinction between normal irregular galaxies and peculiar galaxies, but often galaxies that have been distorted by interactions with other galaxies are termed peculiar. As a result of these interactions, peculiar galaxies tend to have a greater than usual amount of gas and dust, an irregular shape, active star formation, and an unusual composition. More examples of peculiar galaxies are shown in Chapter 9.

11.6 Radio Galaxies

A radio galaxy is a galaxy that emits radiation in the radio wavelengths of the electromagnetic spectrum. If we look at the diagram of the electromagnetic spectrum in Chapter 10, we can see that radio waves, which include microwaves, make up a large part of the spectrum. Because radio waves are not absorbed or reflected by the atmosphere,

Very Large Array (VLA)

Courtesy of NRAO/AUI and NRAO

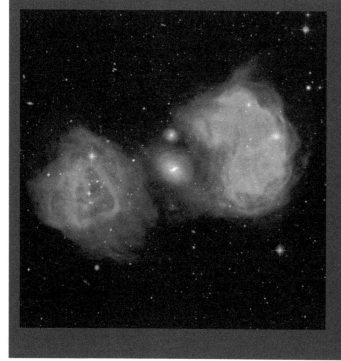

Radio galaxy Fornax A (NGC 1316)

Here we see an optical (visible light) image of the elliptical galaxy Fornax A in the center. Above it is NGC 1317, a small spiral galaxy that is merging with Fornax A. Gas and dust are being pulled from the smaller galaxy into the supermassive black hole at the center of Fornax A, and the black hole is shooting powerful jets of excess matter out into space, forming the two huge reservoirs of hot gas that are emitting radio waves. Each lobe is about 600,000 light years across. The radio waves have been colored orange.

Courtesy of NRAO/AUI and J. M. Uson

clouds, or rain, astronomers can use land-based radio telescopes to detect radio waves coming to Earth from objects in space. Radio waves can be detected day or night because they don't require a dark sky to be visible to radio telescopes. Shorter wavelengths, such as X-rays and gamma rays, are blocked by the atmosphere and need space telescopes for detection. To get more detailed images, radio telescopes can be linked together in an array to act as one big telescope. For example, the Very Large Base Array consists of 10 telescopes

Three views of elliptical radio galaxy Hercules A (3C 348)

Hercules A is 1,000 times as massive as the Milky Way, its black hole (at 2.5 billion times the mass of the Sun) is 1,000 times bigger than Sgr A* and it is one of the brightest sources of radio waves in the sky.

Upper left: We see a visible light image of an ordinary looking Hercules A galaxy. The small white blob near the center of the galaxy is most likely a smaller galaxy that is merging with Hercules A, causing energetic bursts from the supermassive black hole.

Upper right: When radio wave imagery is added, Hercules A becomes a spectacular radio galaxy with two lobes of very high energy plasma beams (subatomic particles and magnetic fields) shooting out from the galactic center at nearly the speed of light. The lobes are 1 million light years long and the ring-like structures in the lobes indicate that there may have been multiple outbursts over time.

Lower right: X-ray imagery has been added to show a giant cloud of multimillion-degree gas. The artist has chosen blue instead of pink for the radio waves.

Image credits — Upper left: NASA, ESA, S. Baum and C. O'Dea (RIT), and the Hubble Heritage Team (STScI/AURA), Upper right: NASA, ESA, S. Baum and C. O'Dea (RIT), R. Perley and W. Cotton (NRAO/AUI/NSF), and the Hubble Heritage Team (STScI/AURA) Lower right: X-ray: NASA/CXC/SAO, Optical: NASA/STScI, Radio: NSF/NRAO/VLA

linked together over a distance of more than 5,000 miles, making a telescope that has sharp enough detail to read a street sign in New York City from Los Angeles.

Most celestial objects emit radio waves, and some galaxies emit especially strong ones. The theory is that these radio galaxies have a very active supermassive black hole at the center. The black hole draws in and consumes matter and then spits out material in explosive, fast moving jets. It is also thought that when the black holes of colliding galaxies get close enough to merge, they generate an immense amount of energy.

11.7 Summary

- The Hubble classification scheme for galaxies divides galaxies into four major groups: spiral, barred spiral, elliptical, and irregular.

- Spiral and barred spiral galaxies both have spiral arms extending from a galactic bulge. In the case of barred spiral galaxies, the galactic bulge is elongated.

- Elliptical galaxies are circular or elongated.

- Irregular galaxies have no defined shape or features.

- Radio galaxies emit very strong radio waves.

11.8 Some Things to Think About

- Why do you think astronomers find it useful to classify galaxies according to their shape?

    What disadvantages do you think there might be to this type of classification system?

- What features would you expect to see in a spiral galaxy?

- Looking at the galaxy images in Section 11.2, what difficulties do you think you might encounter if you were classifying spiral galaxies as Sa, Sb, or Sc?

- Do you think it could be difficult to determine whether a galaxy is a spiral or barred spiral galaxy? Why or why not?

- Do you think there could be times when you could make a good guess that a galaxy is a spiral galaxy but not whether it is a barred spiral galaxy? Why or why not?

Chapter 11: Other Galaxies

- What makes an elliptical galaxy different from a spiral galaxy?
- How do you think dust lanes can help determine the orientation of a galaxy?
- Why do you think irregular galaxies are hard to classify?
- What advantages do radio telescopes have over space telescopes? Over optical telescopes?
- When would you want to use a space telescope rather than a radio telescope?
- How have radio telescopes enhanced our knowledge of the Hercules A and Fornax A galaxies?

Chapter 12 Exploding Stars and Other Stuff

12.1	Introduction	117
12.2	Red Giants, White Dwarfs, and Novae	117
12.3	A White Dwarf Goes Supernova	123
12.4	Supergiant Stars and Supernovae	124
12.5	Summary	128
12.6	Some Things to Think About	128

Crab Nebula image courtesy of NASA, ESA, J. Hester and A. Loll (Arizona State University)

Chapter 12: Exploding Stars and Other Stuff

12.1 Introduction

When we look at the beautiful night sky sparkling with thousands of stars, we don't often think about all the objects we can't see. But by using land-based or space-based telescopes, we can find many different and amazing objects in the sky.

In the previous three chapters we explored our Milky Way Galaxy and other galaxies in the universe. Now we'll take a look at some other interesting features of the universe, such as red giants, white dwarfs, novae, and supernovae.

12.2 Red Giants, White Dwarfs, and Novae

At over 13 billion light years away, galaxy z8_GND_5296 is one of the most distant discovered

Courtesy of NASA, ESA, V. Tilvi (Texas A&M University), S. Finkelstein (University of Texas, Austin), and C. Papovich (Texas A&M University)

Viewing faraway stars is like going back in time. Light from the most distant stars we have been able to observe has traveled billions of light years to get to Earth! No one really knows for sure how stars are born and how they die because star formation and the growth and death of stars happen over periods of millions and billions of years. But with so many stars existing in the universe and because we can study stars that have existed over a time span of billions of years, astronomers can find stars at all different stages of growth and development. Just as a scientist studying humans might look at people at all different stages in life to get an idea of how humans grow and change, astronomers look at stars of many different ages to develop theories about how stars change over time.

In talking about the size of stars and other celestial bodies, astronomers use the term solar mass, a unit of measure in which 1 solar mass is equal to the mass of our Sun. It is estimated that our Sun took tens of millions of years to form and that it is now in the main sequence stage, the period of time when a star is fully formed and stable. The main sequence stage is generally the longest period in the life of a star. It is expected that the length of the Sun's main sequence stage will be about 10 billion years and it is now about halfway through this period.

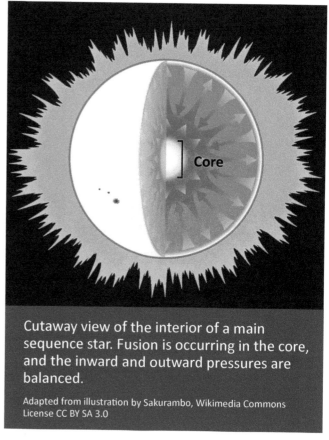

Cutaway view of the interior of a main sequence star. Fusion is occurring in the core, and the inward and outward pressures are balanced.

Adapted from illustration by Sakurambo, Wikimedia Commons License CC BY SA 3.0

A main sequence star uses thermonuclear fusion to fuse hydrogen in its core. Here the hydrogen is under intense pressure from the weight of the matter above it, making the hydrogen extremely hot. The intense heat and gravitational pressure in the core cause the hydrogen nuclei to fuse together in a nuclear reaction that makes helium. Hydrogen fusion occurring within the core releases a massive amount of energy that creates outward pressure in the form of radiation of heat and energy. This outward pressure balances the gravitational force that is pulling matter inward. Because the outward and inward pressures are balanced, the star keeps its shape.

The helium being produced through fusion is heavier than hydrogen, which causes the helium to sink down into the interior of the core. As the star fuses more and more of the core's hydrogen into helium, the amount of helium increases while the amount of hydrogen decreases. When almost all of the hydrogen in the core has been used for fuel and the rate of fusion in the core has slowed drastically, there is not enough outward pressure being created by fusion to balance the inward pressure of gravity. Now the atoms and molecules in the core are forced closer together by gravity, causing the core to contract. This begins a series of events that eventually lead to the death of the star.

The compression of the atoms and molecules in the core releases energy which heats up a thin layer of the hydrogen surrounding the core and causes this layer, or shell, of hydrogen outside the core to fuse. As more hydrogen is burned outside the core, even more of the heavier helium is produced and sinks into the core. At the same time, the core continues to shrink from gravitational pressure. The core becomes denser as more mass is pressed into a small space, the compression causes more heat to be released, and the increase in heat makes hydrogen fusion in the layer outside the core occur at a faster rate.

The star is growing more luminous from the increased fusion and can become as much as 1000 times brighter than the Sun. The energy created by the contraction of the core and the increase in hydrogen fusion is creating outward pressure that causes the star's envelope (the outer non-burning layers) to expand, making the star grow bigger. The expansion of the envelope pushes matter farther from the heat source at the core, causing the outer part of the star to become cooler. The star is becoming unstable and is turning into a big red giant that can eventually become as much as 20 to 100 times the size of the Sun.

The gravitational force acting on the core continues to compress it, making it contract more and more. This causes more heat to be generated until the helium in the core becomes so hot that it can fuse into carbon. Carbon is heavier than helium, so as it is produced, it sinks down into the interior of the core. Now the red giant is developing a non-burning carbon core that is growing denser and

Our Sun as a red giant
It is theorized that in a few billion years our Sun will become a red giant with a diameter of at least 2 AU (a radius of 1 AU). Since Earth's orbit is 1 AU from the Sun, in the red giant stage, the outer edge of the Sun will reach Earth.

Illustration courtesy of Oona Räisänen (User: Mysid), User: Mrsanitazier, Wikimedia Commons, CC BY SA 3.0

getting hotter, surrounded by a helium burning layer that is in turn surrounded by a hydrogen burning layer. Fusion continues to occur at an ever more furious pace with hydrogen fusing into helium which in turn fuses into carbon. As the helium continues to burn, the star begins to run out of fusible fuel and starts to come apart. Stellar winds are created by the radiation from the hot, inner layers of the star, and matter from the envelope is blown out into space. Finally, when all the matter in the envelope has been blasted away, all that is left of the star is the hot, glowing carbon core. The star has become a white dwarf that has no fusible material left in it. It is burned out.

Red giant U Camelopardalis
The star is surrounded by a shell of gas that has been ejected from its envelope by stellar winds.

Courtesy of ESA/Hubble, NASA and H. Olofsson (Onsala Space Observatory)

As the star enters the white dwarf stage, it is surrounded by clouds of gas and dust made up of the matter that was ejected during its red giant stage, The gas and dust clouds form a planetary nebula. Planetary nebulae are among the most beautiful objects to be seen in space and vary greatly in shape and complexity.

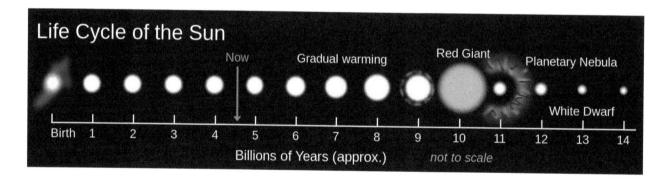

Eventually the gas and dust clouds of the planetary nebula will dissipate, leaving just the naked white dwarf. Because it is the compressed core of a star, a white dwarf is very dense and can have the mass of the entire Sun in a lump about the size of Earth. It is thought that if you could travel to a white dwarf and get a teaspoonful of matter, it would weigh 15 tons on Earth! That would make it really hard to get out of your spacecraft. After billions of years, the white dwarf will have radiated all its heat into space, becoming a cold lump of carbon called a black dwarf.

Chapter 12: Exploding Stars and Other Stuff   121

The term planetary nebula is a bit confusing because a planetary nebula has nothing to do with planets! In the 1700s William Herschel discovered Uranus, and when he observed another round object the color of Uranus, he named it a planetary nebula, thinking it was a planet with rings. Actually, it was a white dwarf surrounded by gas and dust. Unfortunately, the name planetary nebula stuck. Also, the term nebula refers to any hazy or fuzzy-looking celestial object and needs further definition to describe what type of nebula it is.

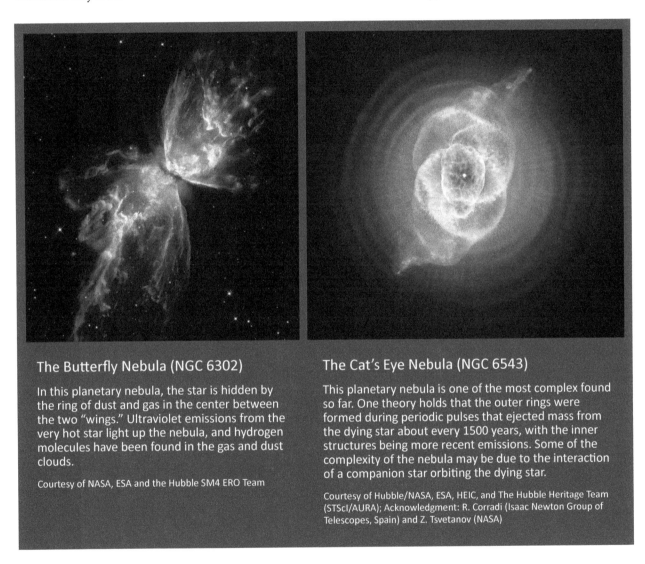

The Butterfly Nebula (NGC 6302)

In this planetary nebula, the star is hidden by the ring of dust and gas in the center between the two "wings." Ultraviolet emissions from the very hot star light up the nebula, and hydrogen molecules have been found in the gas and dust clouds.

Courtesy of NASA, ESA and the Hubble SM4 ERO Team

The Cat's Eye Nebula (NGC 6543)

This planetary nebula is one of the most complex found so far. One theory holds that the outer rings were formed during periodic pulses that ejected mass from the dying star about every 1500 years, with the inner structures being more recent emissions. Some of the complexity of the nebula may be due to the interaction of a companion star orbiting the dying star.

Courtesy of Hubble/NASA, ESA, HEIC, and The Hubble Heritage Team (STScI/AURA); Acknowledgment: R. Corradi (Isaac Newton Group of Telescopes, Spain) and Z. Tsvetanov (NASA)

Some white dwarf stars are part of a binary star system. In a binary star system two nearby stars orbit each other, which is to say they orbit around a common point in space. A white dwarf that is in a binary system can become a nova—a star that has a very sudden, enormous increase in brightness and then fades slowly back to its original luminosity. To fuel this burst of light, gravitational force from the white dwarf pulls hydrogen from the nearby companion star's atmosphere.

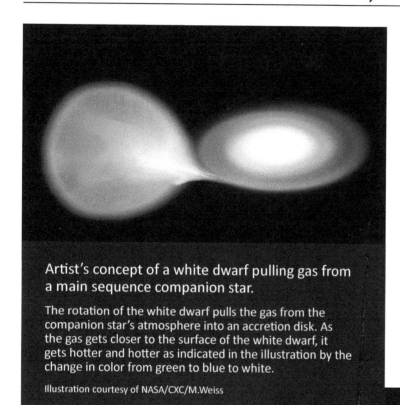

Artist's concept of a white dwarf pulling gas from a main sequence companion star.

The rotation of the white dwarf pulls the gas from the companion star's atmosphere into an accretion disk. As the gas gets closer to the surface of the white dwarf, it gets hotter and hotter as indicated in the illustration by the change in color from green to blue to white.

Illustration courtesy of NASA/CXC/M.Weiss

Because a white dwarf is so dense, its gravitational pull is very strong. The gravitational force and the rotation of the white dwarf pull the stream of gas into orbit around it, creating a layer of matter called an accretion disk. As gas from the accretion disk spirals inward, it falls on the extremely hot surface of the white dwarf and builds up into a layer of gas that gets more and more dense and hotter and hotter. When the hydrogen on the surface of the white dwarf becomes dense and hot enough, there is a sudden thermonuclear explosion with hydrogen fusion occurring at a furious rate. The explosion causes matter to be ejected from the surface of the white dwarf and from the accretion disk. The white dwarf can become a million times brighter than normal as it "goes nova." A nova explosion can release as much as 100,000 times the total energy put out by the Sun in an entire year and can eject matter at velocities of several thousand kilometers per second. The nova is extremely bright for a short period of time and then fades slowly.

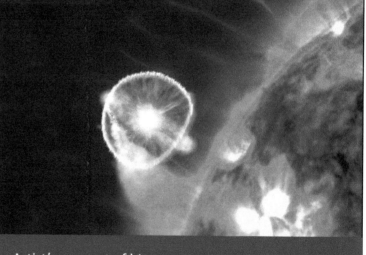

Artist's concept of binary star system V407 Cygni

Here we see a white dwarf gathering gas from its companion red giant star. Gas from the red giant is being pulled into a cloudy-looking accretion disk around the white dwarf. The white dwarf has "gone nova" in the past as shown by the yellow and magenta rimmed circle, or shell, around it. This hot, dense, expanding shell of ejected matter was created by the thermonuclear explosion and is made of high-speed particles, ionized gas, and magnetic fields. The magenta rim represents high-energy gamma rays that are thought to be produced by high-speed particles from the explosion smashing into the red giant's stellar winds.

Courtesy of NASA's Goddard Space Flight Center/S. Wiessinger

A nova can cycle a number of times. The cycle begins as the white dwarf takes hydrogen from the companion star and pulls it into an accretion disk. Then enough hydrogen builds up on the surface of the white dwarf for a thermonuclear explosion to occur, causing a short, intense period of fusion and ejection of matter. The white dwarf stays intact during this process of "going nova." The cycle begins again as the white dwarf once more starts pulling hydrogen from its companion star.

The word nova comes from Latin and means "new." White dwarfs are too faint to be easily detected, and early astronomers could only see these stars when they were bright novae. To these astronomers novae seemed to be newly formed stars.

12.3 A White Dwarf Goes Supernova

Although a supernova is a star whose intensity suddenly increases and then fades slowly, it is different from a nova. A supernova is a star that explodes into stellar debris.

Under the right conditions, a white dwarf star can "go supernova." Recall that a white dwarf is the carbon core of a burned-out, smaller size star. In a binary system a white dwarf can pull hydrogen from a companion star. If a white dwarf in a binary system accretes too much matter from its companion star, the lump of carbon begins to collapse inward from the great gravitational pressure of all that extra mass. This causes the temperature inside the white dwarf to rise dramatically to the point where carbon begins to fuse into heavier elements. Fusion occurs

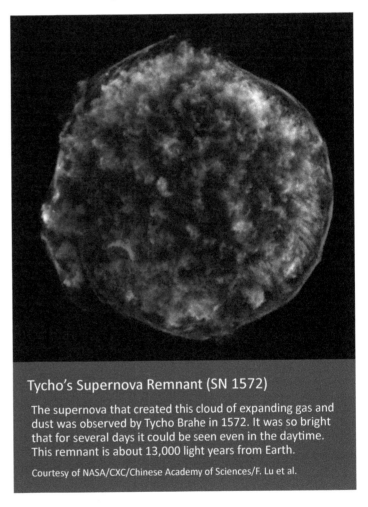

Tycho's Supernova Remnant (SN 1572)

The supernova that created this cloud of expanding gas and dust was observed by Tycho Brahe in 1572. It was so bright that for several days it could be seen even in the daytime. This remnant is about 13,000 light years from Earth.

Courtesy of NASA/CXC/Chinese Academy of Sciences/F. Lu et al.

almost everywhere at once, causing the white dwarf to explode. The expanding cloud of gas and dust ejected by the explosion is called a supernova remnant.

It is predicted that a supernova remnant will last for several hundred thousand years, gradually thinning out as it disperses into the surrounding interstellar medium, the matter that exists in between the stars.

12.4 Supergiant Stars and Supernovae

Supergiant stars are those that are 8 or more solar masses in size. Supergiants have a life cycle that begins in a way that is similar to that of low to medium mass stars, but because supergiant stars have way more mass than smaller stars, the life cycle of a supergiant has additional stages. Like a low to medium mass star, a supergiant in the main sequence stage fuses hydrogen into helium in the core. Also like a low to medium mass star, when the hydrogen in the core is used up, the supergiant star begins fusing hydrogen in the layer around the core and then begins to fuse helium into carbon within the core. The star's envelope begins to expand. It is becoming a red supergiant star. The expansion of the matter in the envelope causes the temperature at the surface to get cooler at the same time as the compression of the core makes the core continue to get hotter. Throughout the red supergiant stage the star stays about the same luminosity although its radius increases greatly.

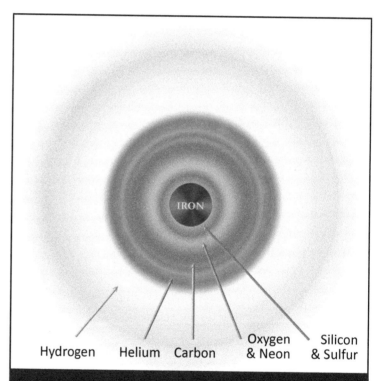

Illustration of the layers in the core of a red supergiant star just before it goes supernova (not to scale).

Starting with the outer layer of hydrogen, as each layer of gas gets closer to the core, it contains a heavier element. As each heavier element is produced it sinks to the layer below it where it is fused to make the next heavier element until, finally, non-fusible iron fills the core. The outer layer of hydrogen is fusing to helium. The helium is fusing to carbon, carbon to oxygen, and so on through the layers down to the iron core.

Illustration courtesy of NASA/CXC/S. Lee

For a red supergiant the process of helium fusing into carbon in the core is just the beginning. Once most of the helium in the core has fused, the pressure from the weight of the great mass in the envelope of the star causes the core, now mostly carbon, to further collapse. This compression generates enough heat and pressure to make carbon in the core begin to fuse into oxygen. The oxygen produced is heavier than carbon and sinks toward the interior of the core. When almost all the carbon has fused into oxygen, the process continues with increased core contraction followed by increased heat and fusion. This cycle repeats through several more stages. The interior of the star becomes layered, with hydrogen burning in the layer outside the core and each layer below the hydrogen containing an increasingly heavier element. Each layer of fusion provides fuel for the layer below it. Carbon fuses to oxygen, oxygen to neon, neon to silicon, and silicon to iron.

In the final stage, the inner core is made of iron, which won't fuse. But still the inward pressure increases. The star is in big trouble. The core gets as compressed as it can possibly get, but the envelope of the star continues to press down harder on the core from the intense gravity, creating even more pressure. Finally, the pressure is so great that the core suddenly collapses inward, or implodes, and turns into an extremely dense neutron star or, for the more massive supergiants, a stellar black hole (a black hole made from a star). The energy created from the collapse of the core into a neutron star or black hole causes a violent rebound expansion that blows the star apart. This is a core-collapse, or Type II, supernova. The collapse of the core to the supernova explosion happens in less than a second!

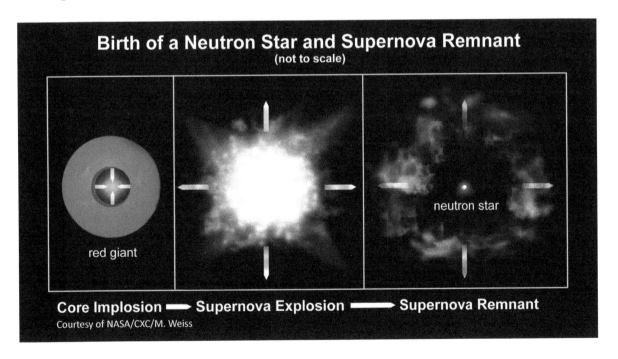

The elements that were created during the different stages of fusion are hurled into space by the supernova explosion at speeds of more than 50 million km per hour. In addition, the tremendous heat and pressure produced by the explosion cause the creation of elements heavier than iron, such as calcium, silver, gold, and uranium. All of these elements are flung out into space, forming rapidly moving clouds of gas and dust that create beautiful supernova remnants.

The explosion of a Type II supernova can be a million times brighter than a nova and billions of times brighter than the Sun. It can even be brighter than the entire galaxy it is found in. Maximum brightness can occur within hours of the beginning of the supernova explosion, and in the few months between the explosion and its fading away, it puts out about as much electromagnetic radiation as the Sun will in its entire lifetime of billions of years.

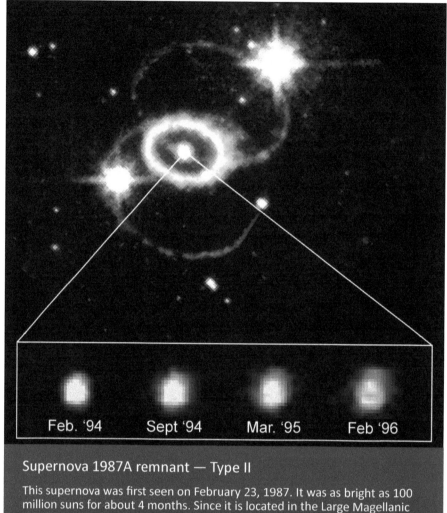

Supernova 1987A remnant — Type II

This supernova was first seen on February 23, 1987. It was as bright as 100 million suns for about 4 months. Since it is located in the Large Magellanic Cloud about 160,000 light years away, the explosion would have occurred about 158,000 BCE. The bright ring around the center is matter ejected from the dying star thousands of years before the supernova explosion. The panel of four white blobs at the bottom of the image show the expansion of the explosion debris which is traveling at nearly 10 million km per hour. The blob on the far right begins to show that the debris is traveling away from the center in two directions, leaving a dim area in the center.

Courtesy of Chun Shing Jason Pun (NASA/ESA/GSFC), Robert P. Kirshner (Harvard-Smithsonian Center for Astrophysics), and NASA/ESA

Chapter 12: Exploding Stars and Other Stuff

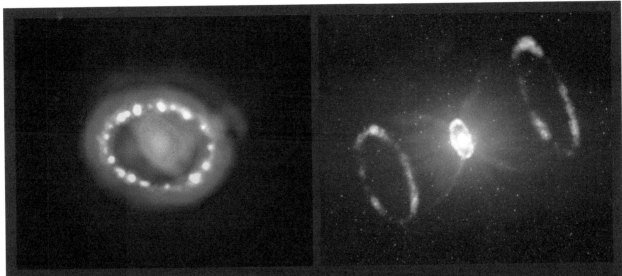

More about Supernova 1987A

Because the explosion of supernova 1987 was observed during recent times, astronomers have been able to study changes it is undergoing as they occur.

Left: A composite image of the inner structure of the supernova remnant. The red is radio waves that show newly formed dust. The green and blue show the expanding shock wave smashing into the ring of older ejecta. This collision is releasing energy that is causing the ring of matter to brighten.

Courtesy of ALMA (ESO/NAOJ/NRAO)/A. Angelich. Visible light image: the NASA/ESA Hubble Space Telescope. X-Ray image: The NASA Chandra X-Ray Observatory

Right: This illustration is based on a theory about how the two red rings seen in the image on the previous page might have formed. Here we are imagining that we are looking at the supernova from a different angle. We can see that the rings are on opposite sides of the center of the explosion rather than being overlapping formations. This indicates that the force of the explosion was directional rather than moving out in a sphere around the star.

Illustration courtesy of ESO/L. Calçada

Type II supernova remnant N49

Courtesy of NASA and The Hubble Heritage Team (STScI/AURA) Acknowledgement: Y. H. Chu (UIUC), S. Kulkarni (Caltech), and R. Rothschild (UCSD)

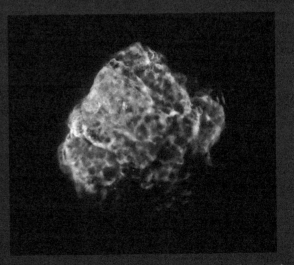

Type II supernova remnant Puppis A (IC 443)

Courtesy of NASA/CXC/IAFE/G. Dubner et al. & ESA/XMM-Newton

Stars are gigantic factories that produce elements by thermonuclear fusion. Stars in the red giant, red supergiant, and nova stages periodically shed matter from their envelope, sending clouds of gas and dust shooting into space. Even more elements are produced and hurled into space when stars explode as supernovae. As these clouds of matter speed through the universe, they run into other clouds of matter from other stars. Eventually, enough matter and energy accumulate for stars, planets, and other celestial objects to form. Our Sun and Earth were formed from stars in this way. And we are all made of elements that come from the Earth and from the energy that is produced by the Sun. We are all made from stars!

12.5 Summary

- A red giant is a low to medium mass star that becomes larger, brighter, and hotter as it begins to burn helium for fuel.

- A white dwarf is the dense core of a less massive star that has used up all of its own fuel.

- A nova is a white dwarf that has pulled hydrogen from a nearby companion star, resulting in a sudden, enormous increase in brightness and then a fading to its original luminosity.

- A supernova is a star that suddenly increases in luminosity as it explodes, and then it slowly fades.

12.6 Some Things to Think About

- When is a star a main sequence star?

- Briefly explain how a main sequence star becomes a red giant.

- How does a nova occur?

- What is the difference between a nova and a supernova?

- What do you find most surprising about supernovae?

- Why do you think scientist Carl Sagan said that we are all made of starstuff?

- Think about what you have learned about astronomy. Are there any theories that you think may change over time? Why?

- Which aspect of astronomy would you most like to study?

Chapter 12: Exploding Stars and Other Stuff

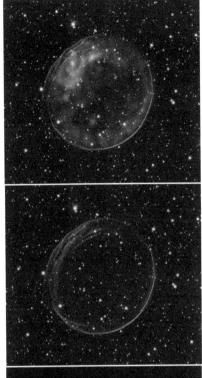

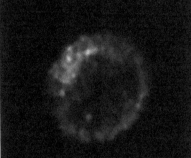

Supernova remnant

SNR 0509-67.5 is about 160,000 light years away in the Large Magellanic Cloud and is 23 light years across. It is expanding at over 11 million mph.

Top: Composite image

Middle: Optical image. Shows the shock wave created by ejecta colliding with interstellar gas. The pink color is glowing hydrogen.

Bottom: X-ray image. Shows ejected matter that has been heated to millions of degrees.

Image credits — Optical: NASA/ESA/Hubble Heritage Team (STScI/AURA), X-ray: NASA/CXC/SAO/J.Hughes et al.

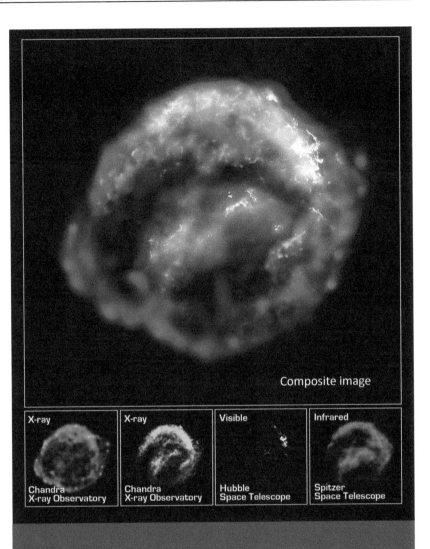

Kepler's Supernova Remnant (SN 1604)

The supernova explosion was seen in 1604 by astronomer Johannes Kepler and others, and was brighter than Jupiter. The supernova remnant is 13,000-23,000 light years away, 14 light years across, and expanding at 4 million mph.

Blue: X-rays show regions of very hot gas and extremely high-energy particles produced by the shock wave from the explosion.

Green: Lower energy X-rays (longer wavelength) indicate cooler gas and materials that were ejected by the explosion of the star.

Yellow: Visible light image shows the shock wave slamming into dense regions of interstellar gas. The bright blobs reveal dense clumps of gas. The filaments are created by the shock wave colliding with lower density interstellar material.

Red: Infrared imaging of microscopic dust particles that have been heated by the shock wave.

Courtesy of NASA/ESA/R. Sankrit and W. Blair (Johns Hopkins University)

Glossary-Index

[Pronunciation Key at end]

Abell 1689 • a massive galaxy supercluster containing many globular clusters, 83

absolute magnitude • see magnitude, absolute

accretion (ə-'krē-shən) **disk** • a layer of matter orbiting a celestial body, 122-123

Alpha Centauri A ('al-fə sen-'tô-rē ā) • a star that is similar to our Sun and is in the Alpha Centauri system; one of the stars closest to our solar system, 70

Alpha Centauri B ('al-fə sen-'tô-rē bē) • a star that is similar to our Sun and is in the Alpha Centauri system; one of the stars closest to our solar system, 70

Alpha Centauri system ('al-fə sen-'tô-rē 'sis-təm) • a triple-star system that contains the stars closest to our solar system—Alpha Centauri A, Alpha Centauri B, and Proxima Centauri, 69-70

Andromeda Galaxy (an-'drä-mə-də 'ga-lək-sē) • the nearest spiral galaxy to our Milky Way Galaxy; found in the constellation Andromeda; also called Messier 31, M31, or NGC 224, 81, 105, 106

apparent magnitude • see magnitude, apparent

apparent solar time • see time, apparent solar

Arches Cluster • the densest globular cluster in the Milky Way Galaxy, 96

Argelander, Friedrich • (1799-1875 CE) German astronomer; published a star catalog of the locations of 325,000 stars, 49

Aristarchus of Samos (a-rə-'stär-kəs of 'sā-mäs) • (310-230 BC [BCE]) Greek astronomer and mathematician; believed in a heliocentric cosmos, 5-6

Aristotle ('a-rə-stä-təl) • Greek philosopher (384-322 BC [BCE]); studied motion and other physical laws; believed in a geocentric cosmos, 5-6

arms • see spiral arms

Arp, Halton C. • (1927-2013 CE) an American astrophysicist who studied and cataloged unusually shaped galaxies, 87

asterism ('as-tə-ri-zəm) • a group of stars that is smaller than a constellation and may be part of a constellation, 50

asteroid ('as-tə-roid) • [*aster*, Gr., star] a small, irregularly shaped celestial body made mostly of rock and minerals, 15, 62-64, 86, 96

Asteroid ('as-tə-roid) **Belt** • a region between Mars and Jupiter that contains numerous asteroids, 62-63

Asteroid Gaspra ('as-tə-roid 'gä-sprə) • an asteroid with an elongated body, 63

Asteroid Kleopatra ('as-tə-roid klē-ə-'pa-trə) • a dog bone shaped asteroid, 63

Asteroid Lutetia ('as-tə-roid lü-'tē-shə) • a very large asteroid that is 100 km (62 mi.) in diameter, 62

Asteroid Vesta ('as-tə-roid 've-stə) • a large asteroid about 525 kilometers (326 mi.) in diameter, 62

astronomer (ə-'strä-nə-mər) • [*aster*, Gr., star; *nomas*, Gr., to assign, distribute, or arrange] a scientist who assigns names to the celestial bodies, including stars, and studies how they exist and move in space, 2

astronomical clock • see clock, astronomical

astronomical (as-trə-'nä-mi-kəl) **unit (AU)** • a unit of measure used by astronomers; one AU is equal to 149,597,870.7 kilometers or 92,955,801 miles, the distance of the Earth from the Sun, 59-60, 81

astronomy (ə-'strä-nə-mē) • [*aster*, Gr., star; *nomas*, Gr., to assign, distribute, or arrange] the study of stars and other objects in space, 2

atmosphere ('at-mə-sfir) • [*atmos*, Gr., vapor] the gaseous layer surrounding a celestial body; the air surrounding the Earth, 12, 15, 26, 31, 38, 40, 63, 66, 75, 111, 112, 120

atmospheric turbulence (at-mə-'sfir-ik 'tər-byə-ləns) • a disturbance of particles in the atmosphere, 12

atom ('a-təm) • one of over 100 fundamental units that make up all matter; composed of protons, neutrons, and electrons; an element (see also *Focus On Middle School Chemistry Student Textbook*), 34, 35

AU • see astronomical unit

aurora (ə-'rôr-ə) • streamers or arches of light that appear in the upper atmosphere when the Earth's magnetic field traps particles that have been charged by a solar storm; also called northern and southern lights, 26

axis ('ak-səs) [plural, **axes** ('ak-sēz)] • an imaginary straight line around which an object rotates, 21-22, 24, 31, 52, 53-54, 61

Barnard's ('bär-nərdz) **star** • the next nearest star to our solar system after the stars in the Alpha Centauri system, 70

barred spiral galaxy • see galaxy, barred spiral

basalt (bə-'sôlt) • a dense, fine-grained igneous rock formed from hardened lava from volcanoes, 30

Glossary-Index

belt • in astronomy, a dark band on a gaseous planet; believed to be made of lower and warmer clouds of gas, 41, 42, 43, 44

Belt, Kuiper • see Kuiper Belt

binary star system • see star system, binary

black dwarf star • see star, black dwarf

black hole • a section in space where, due to extremely high gravity, no light can escape; thought to occur when a large amount of mass from a stellar remnant occupies a small amount of space, 84-85, 96, 109, 114, 125

black hole, stellar • a black hole created as the result of a very massive supergiant star going supernova, 125

black hole, supermassive • a black hole that has a mass greater than 1 million Suns combined, 84, 85, 96, 109, 114

breccia ('bre-chē-ə) • a type of rock formed from soil and pieces of rock squeezed together under high heat and pressure, 30-31

bulge, central • see galactic bulge

burning • in astronomy, the process of thermonuclear fusion in a star, 118, 119, 124

calendar chart • a chart that lists the stars visible from a location on Earth during a particular month, 51

calorie ('ka-lə-rē) • a unit of measurement of energy, 35-36

Canis Major ('kā-nəs 'mā-jər) • "The Big Dog," a constellation that contains VY CMa, the largest visible star, 71

cartography (kär-'tä-grə-fē) • the art and science of making maps, 48

cartography, celestial (kär-'tä-grə-fē, sə-'les-chəl) • the art and science of mapping celestial bodies; also called uranography, 48

celestial (sə-'les-chəl) • having to do with outer space, 2

celestial (sə-'les-chəl) **body** • an object that exists in space, 2

celestial cartography • see cartography, celestial

celestial clock • see clock, celestial

central bulge • see galactic bulge

Cepheid star • see star, Cepheid

charge, electric (i-'lek-trik) • a negative or positive charge on a particle resulting from the gain or loss of electrons; (see *Focus On Middle School Physics Student Textbook*), 26, 34, 72

charge, negative • when an atom has gained an electron and thus has more electrons than protons, it has a negative charge (see also *Focus On Middle School Physics Student Textbook*), 34

charge, positive • when an atom has lost an electron and thus has more protons than electrons, it has a positive charge (see also *Focus On Middle School Physics Student Textbook*), 34

charged particles • particles that carry an electric charge (see also *Focus On Middle School Physics Student Textbook*), 26, 34, 72

chemistry ('ke-mə-strē) • the field of science that studies the composition, structure and properties of matter, 4, 34

Circumstellar Habitable Zone (sər-kəm-'ste-lər 'ha-bə-tə-bəl 'zōn) • the area of a solar system where an Earth-like planet would be at the right temperature to be suited for life as we know it, 74-76

clock, astronomical (as-trə-'nä-mi-kəl) • an instrument that provides information about astronomical movements of celestial bodies and keeps track of time; also called a celestial clock, 55-56

clock, celestial (sə-'les-chəl) • an astronomical clock, 55-56

cold planet • see planet, cold

comet ('cä-mət) • a celestial body that is a large chunk of ice and dirt, 15, 64-65, 66, 86

Comet, Hale-Bopp • see Hale-Bopp Comet

Comet, Halley's • see Halley's Comet

compound telescope • see telescope, compound

compress (kəm-'pres) • to squeeze into a smaller space, 34, 84, 118, 119, 123, 124

constellation (kän-stə-'lā-shən) • a group of stars that forms a pattern in the sky, 3, 48-51, 71

Copernicus, Nicolaus (kō-'pər-ni-kəs, 'ni-kō-läs) • (1473-1543 CE) the astronomer who reintroduced the idea of a heliocentric cosmos, 6

core • the inner-most layer of Earth, the Moon, the Sun, and other celestial bodies; for Earth and the Moon —believed to be made of iron and nickel, 32, 85, 117-119, 122, 123-124

core-collapse supernova • see supernova, core-collapse

cosmos ('käz-mōs) • our solar system; more broadly, the orderly, harmonious system that includes everything that exists in space — the universe, 4-6

crater ('krā-tər) • a bowl-shaped depression around the opening of a volcano; a hole in the ground from a meteorite impact, 32, 39, 41, 63

crust • the outer shell of the Earth or the Moon; made of rocks and soil, 32

dwarf planet • see planet, dwarf

$E = mc^2$ • Einstein's famous equation that shows that mass can be converted to energy, 35-36

Earth • the planet we live on, 5-6, 12, 14, 20-27, 33, 38, 39, 40, *45*, 52-54, 59-61, 63, 64-66

earthquake • a shaking of the earth caused by the movement of pieces of the crust; motion of the crust of the Moon, 32

eclipse (i-'klips) • the obscuring of a celestial body when another celestial body moves between it and the observer, 20, 27

eclipse, lunar (i-'klips, 'lü-nər) • a darkening of the Moon as the Moon passes behind the Earth and the Earth's shadow falls on the Moon, 20, 27

eclipse, solar (i-'klips, 'sō-lər) • a darkening of the Sun when the Moon passes between the Earth and the Sun and the Sun's rays are blocked from reaching Earth, 27

Einstein ('īn-stīn), **Albert** • (1879-1955 CE) regarded as one of the most influential scientists of all time; the father of modern physics, 35

electric (i-'lek-trik) **field** • a region that is electrically charged (see also *Focus On Middle School Physics Student Textbook*), 4

electromagnetic radiation (i-lek-trō-mag-'ne-tik rā-dē-'ā-shən) • electromagnetic waves that are emitted or radiated, 100, 126

electromagnetic spectrum (i-lek-trō-mag-'ne-tik 'spek-trəm) • the full range of electromagnetic waves from short gamma rays through long radio waves, 99-100, 112

electromagnetic (i-lek-trō-mag-'ne-tik) **wave** • a combination of electric and magnetic fields, 85

electron (i-'lek-trän) • one of the three fundamental particles that make up atoms; has almost no mass compared to protons and neutrons; carries a negative electric charge; electrons form the bonds between atoms in molecules (see also *Focus On Middle School Chemistry Student Textbook*), 34

element ('e-lə-mənt) • an atom; basic unit of matter (see also *Focus On Middle School Chemistry Student Textbook*), 30, 33, 122, 124, 125, 127

elliptical (i-'lip-ti-kəl) • not fully circular, 52, 60, 87

elliptical galaxy • see galaxy, elliptical

emission nebula • see nebula, emission

envelope ('en-və-lōp) • in astronomy, the outer non-burning layers of a star, 119, 120, 124, 125, 128

equator (i-'kwā-tər) • the imaginary line around the Earth or other planet that divides it into a north half (northern hemisphere) and a south half (southern hemisphere), 21, 22, 41, 42, 52

exoplanet ('ek-sō-pla-nət) • a planet that orbits a star outside our solar system; also called an extrasolar planet 72-76

extrasolar planet • see exoplanet

false-color images • images made from wavelengths outside the visible light spectral band that have colors assigned by artists and computers, 100, 101, 102

first quarter moon • see moon, first quarter

focal ('fō-kəl) **length** • the distance from a lens to its focal point, 11

focal ('fō-kəl) **point** • the spot at which rays of light entering a lens come together and produce an image, 10, 11

full moon • see moon, full

fuse ('fyüz) • to combine thoroughly, 34, 35, 117, 118, 119, 122, 123, 124

fusion • see hydrogen fusion, thermonuclear fusion

galactic (gə-'lak-tic) • having to do with a galaxy, 83, 105

galactic (gə-'lak-tic) **bulge** • a large, tightly packed group of stars that is the central part of a galaxy; also called the central bulge, 95-96, 106, 107, 108, 109, 110

galactic (gə-'lak-tic) **center** • the central point around which a galaxy rotates, 84-85, 94, 95, 96, 98, 109, 111

galactic (gə-'lak-tic) **disk** • the thin and thick disks of a galaxy together, 97, 106

galactic (gə-'lak-tic) **halo** • a spherical group of old stars and gas that surrounds a galaxy, 95, 97, 106

galaxy ('ga-lək-sē) • a large collection of stars, gas, dust, planets, and other objects held together by its own gravity, 73, 79-114

galaxy, barred spiral • similar to a spiral galaxy, but the spiral arms originate at the ends of a bar-shaped region that goes through the center of the galactic bulge, 94, 105, 106, 107-108

galaxy cluster • a group of galaxies, 79, 83, 86, 105

galaxy, elliptical (i-'lip-ti-kəl) • an ellipse shaped congregate of stars with the largest density of stars in the middle of the galaxy; does not have arms or a galactic bulge, 82, 105, 106, 109-1110, 112

galaxy, irregular (Irr) • a galaxy that does not fit into one of the other categories, 82, 106, 110-112

Galaxy, Milky Way • see Milky Way Galaxy

galaxy, normal irregular (Irr I) • an irregular galaxy that is not considered to be distorted or misshapen, 111

galaxy, peculiar (Irr II) • a type of irregular galaxy, 86, 111-112

galaxy, radio • a galaxy that emits radiation in the radio wavelengths of the electromagnetic spectrum, 112-114

galaxy, S • in the Hubble classification system, designates a spiral galaxy, 105

galaxy, Sa • in the Hubble classification system, designates a spiral galaxy with a large galactic bulge and tightly wound spiral arms, 106

galaxy, SB ('ga-lək-sē, 'es 'bē) • in the Hubble classification system, the designation for a barred spiral galaxy, 105

galaxy, SBb • in the Hubble classification system, the designation for a barred spiral galaxy with a medium size galactic bulge, 108

galaxy, Sc • in the Hubble classification system, designates a spiral galaxy with a small galactic bulge and more loosely wound spiral arms, 106

galaxy, spiral • a galaxy that has a galactic bulge with spiral arms coming out from it making it look similar to a pinwheel, 82, 94, 105, 106-107

galaxy supercluster • a very large group of galaxies, 83

Galileo Galilei (ga-lə-'lā-ō gal-ə-'lā) • (1564-1642) an Italian scientist considered to be the first modern astronomer, 4, 6, 10

gamma ray • a part of the electromagnetic spectrum that is not visible to the unaided eye, 12, 84, *98*, 99, 112

gas • a substance whose molecules are widely separated from each other as in air or water vapor, 33

Gaspra • see Asteroid Gaspra

Gemini ('je-mə-nē, -nī) **Planet Imager (GPI)** • an instrument that images planets directly from an Earth-based observatory, 74

geocentric (jē-ō-'sen-trik) • [*ge*, Gr., earth or land; *kentron*, Gr., point or center] having the Earth as the center, 5, 6

globular cluster ('glä-byə-lər 'kləs-tər) • a dense, globe-shaped clump of hundreds of thousands of very old stars, 83-84, 96, 97, 98, 106

gravitational (gra-və-'tā-shə-nəl) **force** • the force exerted by objects on one another—the Earth's gravitational force keeps objects on its surface, 23-24, 31, 79, 117, 118, 120, 121

gravity ('gra-və-tē) • the force exerted by objects on one another (see also *Focus On Middle School Physics Student Textbook*), 4, 21, 31, 45, 60, 64, 70, 84, 85, 91, 118, 124

Great Red Spot • a huge long-lived storm on Jupiter, 42, 44

Habitable Zone, Circumstellar • see Circumstellar Habitable Zone

Hale-Bopp Comet • a comet that passed close to Earth in 1997, 64

Halley's Comet • a comet that in 1986 passed close enough to Earth for spacecraft to gather information about it, 64

heliocentric (hē-lē-ō-'sen-trik) • [*helios*, Gr., sun; *kentron*, Gr., point or center] having the Sun as the center, 5, 6

heliocentric cosmos (hē-lē-ō-'sen-trik 'käz-mōs) • sun-centered solar system; proposed by Aristarchus of Samos and confirmed by Galileo, 5, 6

helium ('hē-lē-əm) • a gas with 2 protons, 2 neutrons, and 2 electrons; its symbol is He; (see also *Focus On Middle School Chemistry Student Textbook*), 33-35, 38, 42, 43, 117-119, 123

Herschel ('hər-shəl), **Caroline** • (1750-1848 CE) English astronomer who helped her brother William Herschel map the stars, 93

Herschel ('hər-shəl), **William** • (1738-1822 CE) English astronomer who in 1785 presented a paper about the size and shape of our galaxy, 93, 121

Hipparcos (hi-'pär-kōs) • a satellite that gathered data about the stars from 1989-1993; 49

Hipparcos (hi-'pär-kōs) **Star Catalog** • a catalog of 100,000 of the brightest stars accurately mapped by the Hipparcos satellite, 49

hot planet • see planet, hot

Hubble classification scheme • a way of categorizing the various types of galaxies according to their shape; created by Edwin Hubble, 106-112

Hubble, Edwin • (1889-1953 CE) an astronomer who discovered that there are galaxies outside the Milky Way and in 1936 devised a method of classifying galaxies, 81, 105

Hubble Space Telescope • a telescope that is in orbit around the Earth outside the atmosphere, 12, 49, 82, 87

Hubble Tuning Fork • a method of classifying galaxies according to shape; 104-105

hydrogen ('hī-drə-jən) • a gas that is the simplest atom, made up of one proton and one electron; its symbol is H; a hydrogen atom that has lost its electron can be called a proton (see also *Focus On Middle School Chemistry Student Textbook*), 33-36, 38, 42, 43, 117-119, 120-122, 123, 124

hydrogen fusion ('hī-drə-jən 'fyü-zhən) • the fusing of hydrogen nuclei under extremely high temperature and pressure, causing a massive amount of energy to be released; also called thermonuclear fusion, 34-36, 117-118, 121

IAU • see International Astronomical Union

implode (im-'plōd) • to collapse inward suddenly and violently, 125

inertia (in-'ər-shə) • the tendency of things to resist a change in motion (see *Focus On Middle School Physics Student Textbook*), 4

infrared radiation (in-frə-'red rā-dē-'ā-shən) • a part of the electromagnetic spectrum that is not visible to the unaided eye; heat, 12, 74, 94

inner solar system • see solar system, inner

International Astronomical Union (IAU) • a group of professional astronomers from around the world that determines how celestial bodies should be defined and classified, 3, 44, 45, 49

International Space Station (ISS) • a satellite orbiting Earth that was built by space agencies of several different countries and where astronauts from different countries live and do scientific research, 13

interstellar (in-tər-'ste-lər) • located among the stars, 14, 123

interstellar (in-tər-'ste-lər) **medium** • the matter that exists in between the stars, 124

ion ('ī-än) • an atom that has gained or lost one or more electrons, 34

ionization (ī-ə-nə-'zā-shən) • the process of separating the electron(s) from the nucleus of an atom (see also *Focus On Middle School Chemistry Student Textbook*), 34

ionize ('ī-ə-nīz) • to separate the electron(s) from the nucleus of an atom (see also *Focus On Middle School Chemistry Student Textbook*), 34, 64

Irr • designation for an irregular galaxy, 110

Irr I • designation for a normal irregular galaxy, 111

Irr II • designation for a peculiar irregular galaxy, 111

irregular galaxy • see galaxy, irregular

ISS • International Space Station

Jove (jōv) • in Roman mythology, the god of the sky, 41

Jovian planet • see planet, Jovian

Jupiter ('jü-pə-tər) • a large, gaseous planet in the Earth's solar system; also, a massive exoplanet that has Jupiter-like (Jovian) characteristics, 4, 10, 14, 38, 41-42, 43, 44, 59-61, 62, 66, 74

kelvin ('kel-vən) • a unit of measurement of temperature; 1 kelvin = 273.15 degrees Celsius, 33

Kepler, Johannes ('kep-lər, yō-'hä-nəs) • (1571-1630 CE), German astronomer; developed the laws of planetary motion to describe how the planets move around the Sun, 73

Kepler ('kep-lər) **Space Telescope** • launched into an orbit around the Sun in March 2009 with the purpose of finding Earth-size and smaller planets orbiting stars in the Milky Way Galaxy, 13, 73-74, 75

kiloparsec ('ki-lə-pär-sek) **(kpc)** • 1,000 parsecs, 81, 99

Kleopatra • see Asteroid Kleopatra

kpc • see kiloparsec

Kuiper ('kī-per) **Belt** • a belt of small celestial bodies orbiting beyond Neptune, 44

Lalande (lä-'län-dē) **21185** • one of the stars that is nearest to our solar system, 70

lander • a robotic spaceship that can land on the surface of planets or other celestial bodies to capture images and collect data, 14-16, 65, 66

Large Magellanic (ma-jə-'la-nik) **Cloud** • the larger of two small galaxies close to the Milky Way Galaxy, 80, 97

last quarter moon • see moon, last quarter

Leavitt, Henrietta Swan ('le-vit, hen-rē-'e-tə 'swän) • (1868-1921 CE) studied variable stars and discovered the period-luminosity relationship of Cepheid stars, 80-81

light year (ly) • the distance light can travel in one year, or 9.46 trillion kilometers, 81, 83, 97, 98, 99, 109, 117

Lippershey, Hans ('li-pər-shē, 'hänz) • (1570-1619 CE) a Dutch lens maker who in 1608 filed the first patent for a telescope, 10

luminosity (lü-mə-'nä-sə-tē) • the amount of electromagnetic energy radiated by a celestial body, 80, 81, 98, 121, 124

lunar ('lü-nər) • [*luceo*, L., to shine bright] having to do with the Moon, 30

lunar eclipse • see eclipse, lunar

Lutetia • see Asteroid Lutetia

M31 (Messier 31) • see Andromeda Galaxy

M81 • see Messier 81

Magellanic Cloud, Large • see Large Magellanic Cloud

Magellanic Cloud, Small • see Small Magellanic Cloud

magnetic (mag-'ne-tik) **field** • the area affected by magnetic force (see also *Focus On Middle School Physics Student Textbook*); for Earth, comes out from the poles and extends into space, interacting with the Sun, 4, 25, 26, 32

magnitude ('mag-nə-tüd) • a measurement of brightness of a celestial body, 49, 71, 80-81, 98

magnitude, absolute ('mag-nə-tüd, 'ab-sə-lüt) • the actual brightness of a star, 80, 81, 98

magnitude, apparent ('mag-nə-tüd, ə-'per-ənt) • how bright a star appears from Earth, 80, 81, 98

main sequence ('sē-kwəns) **stage** • the period of time when a star is fully formed and stable, 118, 124

mantle ('man-təl) • the layer of a terrestrial planet or moon that is below the crust, 32

mare ('mär-ā), [plural, **maria** ('mär-ē-ə)] • [*marinus*, L., sea] one of the dark areas on the Moon that early astronomers thought were bodies of water but are actually lava flows, 32, 39

maria • see mare

Mariner ('ma-rə-nər) **10** • a space probe that has photographed Mercury, 39

Mars • a terrestrial planet, the fourth from the Sun in our solar system, 15, 16, 17, 38, 39, 40-41, 59, 60, 61, 62

mass • the property that makes matter resist being moved; commonly, the weight of a substance or object, (see *Focus On Middle School Physics Student Textbook*), 4, 21, 23, 24, 31, 33, 35, 72, 84, 85, 95, 108, 118, 119, 122, 123

mass, solar • a unit of measure in which 1 solar mass is equal to the mass of our Sun, 118, 124

mean solar time • see time, mean solar

Mercury ('mər-kyə-rē) • a terrestrial planet; the closest planet to the Sun in our solar system, 38, 39, 45, 59, 60

Messier ('me-sē-ā) **31 (M31)** • see Andromeda Galaxy

Messier ('me-sē-ā) **81 (M81)** • a spiral galaxy found in the constellation Ursa Major, 118

meteor ('mē-tē-ər) • an asteroid that has entered the Earth's atmosphere, 63

meteorite ('mē-tē-ə-rīt) • a meteor that has reached the surface of Earth or another celestial body, 30, 31, 32, 63

Milky Way Galaxy • the barred spiral galaxy that contains our solar system, 73, 79, 80, 81, 82, 92-102, 105, 106, 108

mineral ('min-rəl, 'mi-nə-rəl) • a naturally formed solid substance that is inorganic (does not contain carbon) and has a highly ordered internal structure of atoms *(see Focus On Middle School Geology Student Textbook)*, 30, 62

moon • [*menas*, Gr., month] a celestial body that orbits a planet, 2, 4, 10, 16, 20, 22-24, 27, 30-32, 66, 55, 62, 66

moon, first quarter moon • a phase of the Moon where the Moon looks like a half circle, 22, 23

moon, full • the phase of the Moon during which the Moon looks like a full circle, 22, 23

moon , last quarter • a phase of the Moon where the Moon looks like a half circle, 23

moon, new • the phase of the Moon during which the Moon looks totally dark, 22, 23

moon, waning crescent ('wān-ing 'kre-sənt) • a phase of the Moon where it looks crescent shaped, 23

moon, waning gibbous ('wān-ing 'ji-bəs [or 'gi-bəs]) • a phase of the Moon where the Moon looks convex in shape, 23

moon, waxing crescent ('wak-sing 'kre-sənt) • a phase of the Moon where it looks crescent shaped, 22

moon, waxing gibbous ('wak-sing 'ji-bəs [or 'gi-bəs]) • a phase of the Moon where the Moon looks convex in shape, 22

nebula ('ne-byə-lə) [plural, **nebulae** ('ne-byə-lē)] • [L., cloud] a fuzzy-looking patch of dust, gases, and ionized matter in interstellar space, 49, 51, 81, 97, 120-121

nebula, planetary ('ne-byə-lə, 'pla-nə-ter-ē) • a nebula that occurs when a supernova collapses and ultraviolet radiation ionizes the surrounding area, 120-121

negative charge • see charge, negative

Neptune ('nep-tün) • a Jovian planet, eighth from the Sun in the Earth's solar system; also, a less massive exoplanet that has Jupiter-like (Jovian) characteristics, 14, 38, 41, 43-44, 45, 59, 60, 74

neutron ('nü-trän) • one of the three fundamental particles that make up atoms; it is found in the nucleus and carries no electric charge — it is "neutral" (see also *Focus On Middle School Chemistry Student Textbook*), 34-35

neutron star • see star, neutron

new moon • see moon, new

Newton, Isaac ('nü-tən, 'ī-zik) • (1643-1727 CE) famous British astronomer and mathematician who invented the Newtonian telescope; the founder of physics, 11

Newtonian telescope • see telescope, Newtonian

NGC 224 • see Andromeda Galaxy

Norma Arm • one of the minor spiral arms of the Milky Way Galaxy, 94

North Pole • the northern end of the Earth's axis; the most northern point on Earth, 21, 22

North Star • the star that appears to be directly over the North Pole of the Earth; also called Polaris, 50, 51

Northern Hemisphere ('he-mə-sfir) • the northern half of the Earth, 3, 48, 50

northern lights • an aurora in the Northern Hemisphere; also called aurora borealis, 26

nova ('nō-və) [plural, **novae** ('nō-vē)] • [L., new] a white dwarf star that increases and decreases in brightness as it takes hydrogen from a neighboring star and uses it for thermonuclear fusion, 117, 121-123, 128

nuclear reaction ('nü-klē-ər rē-'ak-shən) • occurs when the protons and neutrons of an atom are moved in and out of the nucleus, creating energy and changing the structure of the atom, 35, 85, 117

nucleus ('nü-klē-əs) [plural, **nuclei** ('nü-klē-ī)] • in atoms, the central portion that houses the protons and neutrons, 34

obliquity • see orbital obliquity

ocean tide • see tide, ocean

Opportunity (ä-pər-'tü-nə-tē) • rover that landed on Mars in 2004, 16

optical light • see visible light

orbit ('ôr-bət) • the curved path of a celestial body as it travels around another celestial body, 5, 6, 12, 13, 15, 21, 22, 30, 31, 38, 40- 45, 52-54, 55, 59-61, 64, 65, 69, 70, 72-74, 79, 84, 95, 96, 121, 122

orbital obliquity ('ôr-bə-təl ō-'bli-kwə-tē) • the tilt of the Earth's axis, 21, 24

orbiter ('ôr-bə-tər) • a spacecraft that goes into orbit around celestial bodies other than Earth, 15-16, 65

Orion (ə-'rī-ən) • a constellation that looks like a hunter with a belt, club, and shield; also called Orion the Hunter, 2, 3

Orion (ə-'rī-ən) **Spur** • the small, partial spiral arm where Earth resides in the Milky Way Galaxy, 94

outer solar system • see solar system, outer

parallax ('per-ə-laks) • the effect in which an object appears to change position when viewed from two different positions, for example when viewed with one eye and then the other; for measuring distance to a star, the star is viewed from two different points in the Earth's orbit allowing for calculations to be made, 79

parsec (pc) ('pär-sec) • a unit of measure used by astronomers; equals 206,260 AUs, 70, 71, 81, 99

pc • abbreviation for parsec

peculiar galaxy • see galaxy, peculiar

period • in variable stars, the length of time it takes a star to go from maximum luminosity through minimum luminosity and back to maximum, 80-81, 98

period-luminosity (lü-mə-'nä-sə-tē) **relationship** • In a variable star, the relationship between the length of the period and the luminosity, 81, 98

Perseus ('pər-süs, 'pər-sē-əs) **Arm** • one of the major spiral arms of the Milky Way Galaxy, 94

phases of the moon • the changes in the Moon's appearance as it is viewed from Earth, 2, 22-23

physics ('fi-ziks) • [*physika*, Gr., physical or natural] the field of science that investigates the basic laws of the natural world, 4, 6, 34-36

planet • [*planetai*, Gr., wanderer] a large spherical celestial body that orbits a sun, has enough mass to have its own gravity, and has cleared its orbit of other celestial bodies, 4-5, 13, 15, 21-22, 25, 38-44, 59-61, 65-66, 69, 72-76, 79, 85, 9

planet, cold • a planet that is far from its sun, 74

planet, dwarf • a celestial body that orbits the Sun and has a spherical shape but is too small to disturb the orbits of other planets, 44-45

planet, extrasolar • see exoplanet

planet, hot • a planet that is close to its sun, 74

planet, Jovian ('jō-vē-ən) • a large planet that is similar to Jupiter; also called a gas giant; Jupiter, Saturn, Uranus, and Neptune, 38, 41-44, 59-60, 61, 74

planet, terrestrial (tə-'res-trē-əl) • a planet that is Earth-like; made of rocky materials and closer to our Sun; Mercury, Venus, Earth and Mars, 38-41, 59-61

planetary nebula • see nebula, planetary

Pluto ('plü-tō) • once considered the 9th planet, is now classified as a dwarf planet or plutoid; this re-classification is in dispute, 38, 44-45

plutoid ('plü-toid) • a celestial body that has enough gravity to have formed a spherical shape, has not cleared its orbit, and exists beyond the orbit of Neptune, 45

Polaris (pə-'ler-əs) • the star that appears to be directly over the North Pole of the Earth; also called the North Star, 50-51

positive charge • see charge, positive

probe • see space probe

proton ('prō-tän) • one of the three fundamental particles that make up atoms; carries a positive electric charge (see also *Focus On Middle School Chemistry Student Textbook*), 34-35

proton, hydrogen • see hydrogen

protostar • the early stage of a star that is being formed, 86

Glossary-Index

Proxima Centauri ('präk-sə-mə sen-'tô-rē) • the star that is nearest to our solar system; part of the Alpha Centauri system, 70

pulsate ('pəl-sāt) • to move rhythmically; to increase and decrease, 80

radiate ('rā-dē-āt) • to send out rays such as electromagnetic waves; to emit, 33, 100, 120

radio galaxy • see galaxy, radio

radio waves • the longest wavelengths in the electromagnetic spectrum, 94, 99, 111-113

reaction, nuclear • see nuclear reaction

red giant star • see star, red giant

red hypergiant (hī-pər-'jī-ənt) **star** • see star, red hypergiant

red supergiant star • see star, red supergiant

reflector telescope • see telescope, reflector

refractor telescope • see telescope, refractor

Rosetta (rō-'ze-tə) **spacecraft** • orbited Comet 67P/Churyumov-Gerasimenko and sent a lander to the surface of the comet, 65

rotation, true • see true rotation

rover • a robotic spaceship that can land on the surface of a planet or asteroid and then travel around, capturing images and collecting data, 14, 16-17, 41

RR Lyrae star • see star, RR Lyrae

S galaxy • see galaxy, S

Sa galaxy • see galaxy, Sa

Sagittarius A* (sa-jə-'ter-ē-əs ā stär); also called **Sgr A*** (pronounced (Saj ā stär)• the black hole at the center of the Milky Way Galaxy, 85, 96

Sagittarius (sa-jə-'ter-ē-əs) **Arm** • one of the minor spiral arms of the Milky Way Galaxy, 94

satellite ('sa-tə-līt) • a celestial body that orbits a larger celestial body; a machine that is put into orbit around the Earth or another celestial body, 12-14, 22, 49

Saturn ('sa-tərn) • a Jovian planet, sixth from the Sun in the Earth's solar system; has colored rings believed to be made of icy fragments, 14, 16, 38, 41, 42, 59-61, 66

SB galaxy • see galaxy, SB

SBb galaxy • see galaxy, SBb

Sc galaxy • see galaxy, Sc

Scutum-Centaurus Arm ('scü-təm sen-'tô-rəs) • one of the major spiral arms of the Milky Way Galaxy, 94

seismic ('sīz-mik) • relating to an earthquake, 32

seismometer • (sīz-'mä-mə-tər) an instrument that detects motion or vibration in the ground, 32

Shapley, Harlow ('shap-lē, 'här-lō) • (1885-1972 CE) an American astronomer who devised a method to use variable stars to find the center of the Milky Way Galaxy, 98

sidereal (sī-'dir-ē-əl) • [*sidus*, L., star] having to do with stars or constellations, 54

sidereal (sī-'dir-ē-əl) **day** • the length of time it takes for Earth to make one true rotation, 54

sidereal (sī-'dir-ē-əl) **time** • time measured by using Earth's position relative to distant stars rather than relative to the Sun; also called star time, 54

Sirius ('sir-ē-əs) • the brightest star in the sky, 71

Small Magellanic (ma-jə-'la-nik) **Cloud** • the smaller of two small galaxies close to the Milky Way Galaxy, 80, 97

Sojourner (sō-'jərn-ər) • a Mars rover, 16

solar day • The time it takes for the Sun to go from its highest position in the sky on one day to its highest position on the next day, 52-54

solar eclipse •see eclipse, solar

solar mass • see mass, solar

solar ('sō-lər) **storm**, a storm on the Sun, 25, 26

solar system ('sō-lər 'sis-təm) • a group of celestial bodies and the one or more suns they orbit, 4, 6, 31, 38, 43, 44, 59-66, 69-76, 79, 82, 92, 94

solar system ('sō-lər 'sis-təm), **inner** • the four terrestrial planets, which are closer to our Sun than the Jovian planets, 61, 66

solar system ('sō-lər 'sis-təm), **outer** • the four Jovian planets, which are farther from our Sun than the terrestrial planets, 61

solstice ('säl-stəs) • the point in time when the Earth's axis is pointed the most directly at the Sun (summer solstice) or the farthest away from the Sun (winter solstice), 9

South Pole • the southern end of the Earth's axis; the most southern point on Earth, 21, 22

southern lights • an aurora in the Southern Hemisphere; also called aurora australis, 26

space probe • a robotic spaceship that can travel far distances, capturing images and collecting data, 14, 39, 66

spectral ('spek-trəl) **band** • a group of wavelengths in the electromagnetic spectrum, 100

spectral ('spek-trəl) **band, visible** ('vi-zə-bəl) • the group of wavelengths in the electromagnetic spectrum that can be seen with the unaided eye, 100

spectrum ('spek-trəm) • a continuous range, such as the range of wavelengths in the electromagnetic spectrum, 98, 99

spectrum, electromagnetic ('spek-trəm, i-lek-trō-mag-'ne-tik) • the full range of electromagnetic waves from short gamma rays through long radio waves, 98-99, 111

spectrum, visible ('spek-trəm, 'vi-zə-bəl) • the part of the electromagnetic spectrum we can see with our unaided eyes; visible light; optical light; visible spectral band, 98, 99

spiral arms • in a galaxy, groups of stars that extend from the galactic bulge in a pinwheel-like form, 94, 96, 105, 106, 107, 108, 109, 111

spiral galaxy • see galaxy, spiral

Spirit • a Mars rover, 16

star • a celestial body that generates light and heat energy, 33

star atlas • a collection of maps of celestial bodies; a star catalog 49-51

star, black dwarf • a cold lump of carbon that is the remains of a burned out star, 120

star catalog • a collection of maps of celestial bodies; a star atlas 49-51

star, Cepheid ('se-fē-id) • a type of variable star that pulsates, increasing and decreasing in size and brightness, 80-81, 98

star map • a star atlas, 49-51

star, neutron (nü'-trän) • the remains of a supernova; a stellar remnant, 125

star, RR Lyrae ('är 'är 'lī-rē) • a variable star with a short period of somewhere between four hours and one day, 97, 98

star, red giant • a star that has used up its hydrogen during thermonuclear fusion and becomes hotter, bigger and brighter as it burns helium, 117-120, 122, 128

star, red hypergiant (hī-pər-'jī-ənt) • a massive red star that has a diameter of between 100 and 2100 times that of our Sun and may be thousands to millions times brighter than the Sun, 71

star, red supergiant • a supergiant star that is unstable and fusing hydrogen in the layer outside its core, causing the envelope to expand, 124-128

star, supergiant • a star that is 8 or more solar masses in size, 124-128

star system, binary ('bī-nə-rē) • two nearby stars that orbit a common point in space, 121-122, 123

star time • sidereal time, 54

star, variable ('ver-ē-ə-bəl) • a star whose luminosity (brightness) changes with time; a pulsating star, 80-81, 98

star, white dwarf • a star that has used up all of its fuel, leaving just its core, 117-128

stellar ('ste-lər) • relating to a star or stars, 71-72

stellar ('ste-lər) **black hole** • see black hole, stellar

stellar ('ste-lər) **disk** • the galactic disk, 97

stellar nursery ('ste-lər 'nərs-rē) • a massive cloud of gas and dust where stars are born within galaxies, 86

stellar ('ste-lər) **outburst** • an eruption of electrically charged particles from a star's surface, 71-72

stellar ('ste-lər) **wind** • a strong emission of radiation from the hot, inner layers of a star and matter from the star's envelope, 120

Stonehenge (stōn'-henj) • an ancient site found in England; made of huge stones set in circular shapes; is thought to have been built as a solar tracking system, 9

Sun • the star around which Earth rotates; made mainly of hydrogen and helium gases; provides the Earth with energy, 21-27, 33-36, 52-54

sundial • a device that records the Sun's daily motion across the sky, 52

supercluster, galaxy • see galaxy supercluster

super-Earth • an exoplanet that has a mass up to ten times that of Earth, 74

supergiant star • see star, supergiant

supermassive black hole • see black hole, supermassive

supernova (sü-pər-'nō-və) [plural, **supernovae** (sü-pər-'nō-vē)] • a star that explodes and becomes stellar debris; its intensity suddenly increases, and then it slowly fades from view, 123-128

supernova, core-collapse • the explosion of a supergiant star caused by the energy created from the collapse of the core into a neutron star or black hole; also called a Type II (tīp tü) supernova, 125-127

supernova remnant (sü-pər-'nō-və 'rem-nənt) • the expanding cloud of gas and dust ejected by the explosion of a star in a supernova, 123-124, 126-128, 129

telescope ('te-lə-skōp) • [tele-, Gr., from afar or far off; skopein, Gr., see, watch, or view] a scientific instrument that uses lenses to magnify distant objects in space, 4, 10-13

telescope, compound ('te-lə-skōp, 'käm-paủnd) • a telescope that combines elements of refractor and reflector telescopes, 10, 11

Telescope, Hubble Space • see Hubble Space Telescope

Telescope, Kepler • see Kepler Space Telescope

Glossary-Index 139

telescope, Newtonian ('te-lə-skōp, nü-'tō-nē-ən) • a common reflector telescope named after Isaac Newton, 11

telescope, reflector ('te-lə-skōp, ri-'flek-tər) • a telescope that uses a combination of mirrors and lenses to focus incoming light and magnify an image, 10, 11

telescope, refractor ('te-lə-skōp, ri-'frak-tər) • a telescope with a lens at one end and an eyepiece at the other; light entering through the lens is bent, magnifying the image, 10

terra ('ter-ə) [plural, **terrae** ('ter-ē)] • [*terra*, L., land] a light area on the Moon's surface that contains rugged craters, 32

terrestrial (tə-'res-trē-əl) • (terra, L., earth) Earth-like; relating to the Earth, 38

terrestrial planet • see planet, terrestrial

thermonuclear fusion (thər-mō-'nü-clē-ər 'fyü-zhən) • the fusing of hydrogen nuclei under extremely high temperature and pressure, causing a massive amount of energy to be released; process used by our Sun to create energy; also called hydrogen fusion, 34-36, 117-125

thick disk • in a galaxy, an area that contains older stars and surrounds the thin disk, 95, 97

thin disk • in a galaxy, an area that contains the majority of stars in the galaxy and orbits the galactic bulge, 95-97

tide, ocean • the rise and fall of ocean levels due to gravitational forces of the Moon and the Sun, 24, 25

time, apparent solar • time measured from the direct observation of the Sun, 52-53

time, mean solar • time based on apparent solar time averaged over the course of a year, 53-54

Tombaugh, Clyde ('täm-bä, clīd) • an astronomer who discovered Pluto in 1930, 44

transit ('tran-sət) • in astronomy, the passage of a smaller celestial body in front of a larger one, 73-74

transit method • a method of detecting planets by observing the tiny dimming of a star's brightness as a planet passes in front of, or transits, the star it is orbiting, 73-74

true rotation • the actual rotation of Earth, 53-54

turbulence, atmospheric • see atmospheric turbulence

Tycho (tē-kō) **Star Catalog** • a catalog of 2 million dimmer stars mapped by the Hipparcos satellite, 49

Type II supernova • see supernova, core-collapse

universe ('yü-nə-vərs) • our solar system; more broadly, everything that exists in space; the cosmos, 4-5

uranography (yùr-ə-'nä-grə-fē) • [*uranos*, Gr., heavens; *graphe*, Gr., to write] the art and science of mapping celestial bodies; also called celestial cartography, 48

Uranus ('yùr-ə-nəs) • a Jovian planet, seventh from the Sun in the Earth's solar system; contains a significant amount of methane, giving it a bluish color; 14, 38, 41, 43-44, *45*, 59-61

vacuum ('va-kyüm) • in astronomy, containing no matter, 31

vaporize ('vā-pə-rīz) • to turn a substance into its gaseous state, 64, 65

variable star • see star, variable

Venus ('vē-nəs) • a terrestrial planet, the second from the Sun in the Earth's solar system, 38, 39, 40, *45*, 59-61

Viking 1 ('vī-king 'wən) • a spacecraft that consisted of both an orbiter and a lander; captured the first image of the surface of Mars, 15

visible ('vi-zə-bəl) **light** • the part of the electromagnetic spectrum we can see with our unaided eyes; the visible spectrum; also called optical light, 74, 95, 99-100

visible spectral ('vi-zə-bəl 'spek-trəl) **band** • visible spectrum, 100

visible spectrum ('vi-zə-bəl 'spek-trəm) • the part of the electromagnetic spectrum we can see with our unaided eyes; visible light; optical light; visible spectral band, 99, 100

Voyager 1 ('vô-ij-ər 'wən) • a space probe launched in 1977 to explore Jupiter and Saturn; is now traveling outside our solar system, 14

Voyager 2 ('vô-ij-ər 'tü) • a space probe that was launched in 1977 to explore Jupiter and Saturn; sent back images of Neptune in 1989; 14, 44

VY Canis Majoris ('vē 'wī 'kā-nəs mə-'jôr-əs) **(VY CMa)** • a red hypergiant star that is the largest star visible in the night sky, 71

waning crescent moon • see moon, waning crescent

waning gibbous moon • see moon, waning gibbous

waxing crescent moon • see moon, waxing crescent

waxing gibbous moon • see moon, waxing gibbous

white dwarf star • see star, white dwarf

Wolf 359 • one of the stars that is nearest to our solar system, 70

x-ray • a part of the electromagnetic spectrum that is not visible to the unaided eye, 84, 94, 95, 98, 112

zone • in astronomy, a light band on a gaseous planet believed to be made of lower and warmer clouds of gas, 41, 42, 43, 44

Pronunciation Key

a	add	j	joy	sh	sure
ā	race	k	cool	t	take
ä	palm	l	love	u	up
â(r)	air	m	move	ü	sue
b	bat	n	nice	v	vase
ch	check	ng	sing	w	way
d	dog	o	odd	y	yarn
e	end	ō	open	z	zebra
ē	tree	ô	jaw	ə	a in above
f	fit	oi	oil		e in sicken
g	go	oo	pool		i in possible
h	hope	p	pit		o in melon
i	it	r	run		u in circus
ī	ice	s	sea		

More REAL SCIENCE-4-KIDS Books
by Rebecca W. Keller, PhD

Building Blocks Series yearlong study program — each Student Textbook has accompanying Laboratory Notebook, Teacher's Manual, Lesson Plan, Study Notebook, Quizzes, and Graphics Package

Exploring Science Book K (Activity Book)
Exploring Science Book 1
Exploring Science Book 2
Exploring Science Book 3
Exploring Science Book 4
Exploring Science Book 5
Exploring Science Book 6
Exploring Science Book 7
Exploring Science Book 8

Focus On Series unit study program — each title has a Student Textbook with accompanying Laboratory Notebook, Teacher's Manual, Lesson Plan, Study Notebook, Quizzes, and Graphics Package

Focus On Elementary Chemistry
Focus On Elementary Biology
Focus On Elementary Physics
Focus On Elementary Geology
Focus On Elementary Astronomy

Focus On Middle School Chemistry
Focus On Middle School Biology
Focus On Middle School Physics
Focus On Middle School Geology
Focus On Middle School Astronomy

Focus On High School Chemistry

Super Simple Science Experiments

21 Super Simple Chemistry Experiments
21 Super Simple Biology Experiments
21 Super Simple Physics Experiments
21 Super Simple Geology Experiments
21 Super Simple Astronomy Experiments
101 Super Simple Science Experiments

Note: A few titles may still be in production.

Gravitas Publications Inc.
www.gravitaspublications.com
www.realscience4kids.com

Printed in the USA
CPSIA information can be obtained
at www.ICGtesting.com
LVHW061230041223
765485LV00031B/1865